Surface Active Chemicals

Surface Active Chemicals

BY

H. E. GARRETT

PERGAMON PRESS

OXFORD · NEW YORK · TORONTO
SYDNEY · BRAUNSCHWEIG

Pergamon Press Ltd., Headington Hill Hall, Oxford
Pergamon Press Inc., Maxwell House, Fairview Park, Elmsford,
New York 10523
Pergamon of Canada Ltd., 207 Queen's Quay West, Toronto 1
Pergamon Press (Aust.) Pty. Ltd., 19a Boundary Street,
Rushcutters Bay, N.S.W. 2011, Australia
Vieweg & Sohn GmbH, Burgplatz 1, Braunschweig

First edition 1972

Library of Congress Catalog Card No. 72-80098

Printed in Hungary

08 0164226

Contents

v

Editors' Preface

WE WERE asked by Sir Robert Robinson, O.M., P.P.R.S., to organize the preparation of a series of monographs as teaching manuals for senior students on the Chemical Industry, having special reference to the United Kingdom, to be published by Pergamon Press. Apart from the proviso that they were not intended to be reference books or dictionaries, the authors were free to develop their subject in the manner which appeared to them to be most appropriate.

The first problem was to define the Chemical Industry. Any manufacture in which a chemical change takes place in the material treated might well be classed as "chemical". This definition was obviously too broad as it would include, for example, the production of coal gas and the extraction of metals from their ores; these are not generally regarded as part of the Chemical Industry. We have used a more restricted but still a very wide definition, following broadly the example set in the special report (now out of print) prepared in 1949 by the Association of British Chemical Manufacturers at the request of the Board of Trade. Within this scope, there will be included monographs on subjects such as coal carbonization products, heavy chemicals, dyestuffs, agricultural chemicals, fine chemicals, medicinal products, explosives, surface active agents, paints and pigments, plastics and man-made fibres.

We wish to acknowledge our indebtedness to Sir Robert Robinson for his wise guidance and to express our sincere appreciation of the encouragement and help which we have received from so many individuals and organizations in the industry, particularly the Association of British Chemical Manufacturers.

The lino-cut used for the covers of this series of monographs was designed and cut by Miss N. J. Somerville West, to whom our thanks are due.

J. DAVIDSON PRATT $\Big\}$ *Editors*
T. F. WEST

CHAPTER 1

General Principles

Introduction

This will seem an odd book at a first glance at the table of contents, from which it will appear to consist of a collection of unrelated topics. Surface activity is almost coextensive with colloid science and hence is exhibited by many classes of compounds and is fundamental to many industries and branches of science. A number of these are the subject of other books in this series, not only because of the size and importance of the industries concerned, but also because the breadth and depth of the scientific and technological problems of each demand separate treatment. These industries include, for example, the dyeing, tanning, and baking industries. Surface activity enters to a large degree into the winning of minerals of economic value, especially in ore flotation processes, and into many biological processes.

The present book treats a number of applications of surface activity on which the writer has worked, not otherwise covered in this series of books. The apparently disparate topics are linked by the common thread of the forces of attraction between molecules in contact. These attractive forces and their consequences relevant to the subsequent technical chapters are discussed in this chapter, in which full treatment is given to aspects of the subject often glossed over or confused in the general literature. The bare minimum of unavoidable discussion of bulk solution phase properties, which is essential for proper understanding of surface phenomena, is included: anything more would have lengthened this already long theoretical chapter out of all proportion.

1

It is hoped that the younger student who may find this theoretical chapter too advanced and too condensed may still gather the general physical picture and that this will serve him in good stead.

Intermolecular Forces

All forces between molecules are electrostatic in origin. For convenience, the van der Waals force of attraction which exists between any two molecules is separated into:

(1) The London, or "dispersion" force, which exists between non-polar atoms and molecules as well as between polar molecules.
(2) The Keesom, or "orientation" force, which exists between any pair of polar molecules, each therefore with permanent dipoles.
(3) The Debye, or "induction" force, which exists between a molecule with a dipole and any other molecule by virtue of the dipole induced in the second molecule by the permanent dipole of the first molecule.

These three together account for the universal van der Waals force of attraction between molecules.

At close contact the strong resistance of the electronic shells of atoms to mutual penetration, or to deformation, leads to a repulsive force which increases very strongly at very short distances.

The average value for the mutual potential energy between two molecules, over all mutual orientations, is given to a first approximation by the expression

$$\bar{E} = \frac{3}{2} \frac{\alpha_1 \alpha_2}{r^6} \frac{I_1 I_2}{I_1 + I_2} - \frac{2\mu_1^2 \mu_2^2}{3r^6 kT} - \frac{\alpha_1 \mu_2^2 + \alpha_2 \mu_1^2}{r^6} + \frac{j}{r^n}, \qquad (1.1)$$

where α_1, α_2 are the polarizabilities of the molecules or atoms, I_1, I_2 are the ionization energies, μ_1, μ_2 are the dipole moments, k is Boltzmann's constant, T is the absolute temperature, r is the distance of separation of the two molecular centres, and j and n are constants.

The first term is always large and often makes the major contribution to $\bar{E}$. Equation (1.1) may be simplified to

$$\bar{E} = j/r^n - m/r^6, \tag{1.2}$$

which, when n is taken as 12, is the Lennard-Jones potential energy expression.

Figure 1.1 shows the repulsive and attractive terms of the Lennard-Jones expression and the sum curve separately plotted. In the sum curve, $\bar{E}$ has a minimum at r_0, which is the distance of closest approach of the two molecules at equilibrium: at distances greater than r_0, all

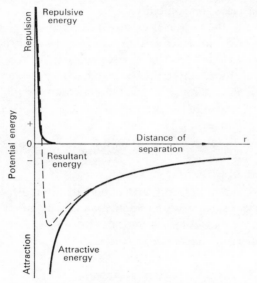

FIG. 1.1. Potential energy between two molecules.

molecules, whatever their nature, attract each other. At distances of separation only very little less than r_0, they repel each other.

In addition to the three components of the van der Waals energy of attraction, there is the Coulombic energy of attraction (or repulsion) between ions of unlike (or like) sign, proportional to the charges on the ions and inversely proportional to the square of the distance of separation, and also ion-dipole and ion-induced dipole energies of attraction.[†]

This simplified discussion applies to atoms and to a close approximation to "small" molecules. For chain molecules of, say, a number of CH_2 groups, in group to group contact, the intermolecular force is to a rough approximation that of the sum of the group–group interactions. Attraction to next neighbour and more distant groups is much less than that of the groups in contact. For large molecules the expression for the interaction becomes much more complicated, but the general principle of universal attraction still holds. The same provisos apply to the section below on solubility.

A corollary to the universal existence of a net energy of attraction between any two molecules is that there is always a net energy of attraction across the boundary surface of contact between any two phases, such as, for example, oil and water. To a good approximation, the van der Waals energy of attraction between two unlike molecules is the geometric mean of the energies of attraction between pairs of the two species of molecules, i.e.

$$\bar{E}_{ab} = 2\sqrt{(\bar{E}_{aa}\bar{E}_{bb})} \tag{1.3}$$

as first suggested by Berthelot in 1898.

This simple fact of universal attraction between molecules is fundamental to every aspect of surface activity and must never be lost sight of. Although oil and water do not mix, **there is no force of repulsion between oil molecules and water molecules.**

Solubility—General

The preceding paragraph raises the question, since there is no force of repulsion between oil and water, Why is oil immiscible with water? Clearly the process of solution must be closely considered.

Consider the introduction by a hypothetical process of a mole of oil from a mass of liquid oil into a large mass of water. Let the two liquids

[†] For detailed discussion of these forces the reader is referred to the standard texts of physical chemistry.

initially be separated by an impermeable partition. Evaporate isothermally one mole of oil from the mass of oil. The process absorbs
the molar latent heat of evaporation λ_{evap}. Expand the vapour reversibly and isothermally by means of a frictionless piston to a volume V_2,
equal to the volume of the mass of water. Work, PV_2, is done against
the external pressure in this process. Transfer the vapour to the space
above the water surface. This is a workless process. Compress the
mole of oil vapour reversibly and isothermally into the water, at the
same time expanding the water so as to create holes of just the right
size and in just the right number to accommodate the incoming oil
molecules. The work of compression of the vapour is $-PV_2$ and hence
cancels with the work of expansion. To expand the water and thus
create holes in its structure, work must be done against the internal
energy of mutual attraction of the water molecules, but part of this
work is recovered from the mutual energy of attraction between the
entering oil molecules and the water molecules surrounding them. The
net work is a function of the difference between the two intermolecular
energies, i.e. $f(\varepsilon_{11} - \varepsilon_{12})$.

The mass of water is to be chosen such that when the mole of oil
has been transferred to the water, the water is just saturated with the
oil. Remove the partition between the phases. This is a workless process. There is now a saturated solution of oil in water in contact
with the pure oil phase. (A similar process could be gone through
to saturate the oil with water: on finally removing the partition
there would then be two mutually saturated phases in contact in
equilibrium.)

In the process considered the heat content of the system has been
increased by $\lambda_{evap} + f(\varepsilon_{11} - \varepsilon_{12})$ and the entropy of the system has been
increased by $\lambda/T + R \ln (V_2/V_1)$, where V_1 is the volume occupied by one
mole of oil at the saturated vapour pressure of the oil, T is the absolute
temperature, and R is the gas constant. Therefore the Gibbs free energy
change for the process is given by

$$\Delta G = \lambda + f(\varepsilon_{11} - \varepsilon_{12}) - T[\lambda/T + R \ln (V_2/V_1)]$$

$$= f(\varepsilon_{11} - \varepsilon_{12}) - RT \ln (V_2/V_1). \tag{1.4}$$

At equilibrium, i.e. for the transfer of a mole of oil to an infinite volume of saturated solution, $\Delta G = 0$. Hence the volume V_2 of the saturated solution containing one mole of oil is determined by the expression

$$RT \ln (V_2/V_1) = f(\varepsilon_{11} - \varepsilon_{12}) \tag{1.5}$$

or

$$V_2 = V_1 \exp [f(\varepsilon_{11} - \varepsilon_{12})/RT]. \tag{1.6}$$

This shows that if the water molecules attract each other much more than they attract oil molecules, V_2 must be large, i.e. the solubility of the oil in water will be small [at constant V_1 the solubility of the oil increases as ε_{12} approaches ε_{11}. At constant $\Delta\varepsilon$ the solubility of the oil decreases with decreasing vapour pressure (increasing V_1)].

Experimentally, for the hydrocarbon parts of the organic molecules dissolved in water, another entropy term which reduces the numerical magnitude of the second term of eqn. (1.4) and hence diminishes the solubility of the organic molecule still further, is found to be necessary. This term has been ascribed by Frank and Evans to the organization of clusters of water around the hydrocarbon chains or rings. These clusters have picturesquely been called "icebergs". This organization of some of the water molecules involves a decrease in the entropy of the system.

The student will realize that the attraction of hydrocarbon for hydrocarbon and of hydrocarbon for water are comparable, whereas the attraction of water for water is much greater: oil does not mix with water **because it is squeezed** out by the powerful forces between water molecules. Similarly, attachment of hydrocarbon chains of long-chain organic compounds solubilized in water by the presence of a water-attracting functional group to, for example, the hydrocarbon portions of proteins, or of the bending of flexible molecules such as soluble heterogeneous polypeptides or proteins in which hydrocarbon side-chains are brought together by the grouping of the amino acids whose water-attracting functions are thus arranged on the outside of the hydrocarbon "clumps", is also a consequence of the much greater water–water attractive force than of the water–hydrocarbon or hydrocarbon–hydrocarbon attractions. Although hydrocarbon–hydrocarbon forces of attraction are also active in such a situation they are not the definitive forces determining the structure, and it is entirely misleading to describe such grouping of hydrocarbon portions of molecules as "hydrophobic bonding". This currently fashionable phrase should be avoided like the plague.

The term V_1 will be a function of the molecular weight and mutual energy of attraction and hence will be a function of the latent heat of evaporation. For solids, V_1 will, similarly, be a function of the latent heat of sublimation equivalent to the sum of the latent heats of melting and of evaporation from the liquid state. In this way the solubilities of solids are related, in part, to the crystal lattice energies. For non-ionizing solids giving approximately ideal solutions, the mole fraction of the solid in the saturated solution is given to a first approximation by the expression

$$\ln N_2 = (\Delta H_m/R)\,(1/T_m - 1/T_s), \tag{1.7}$$

where ΔH_m is the latent heat of melting, T_m is the absolute temperature of melting, and T_s is the temperature of the saturated solution.

In antilog form,

$$N_2 = \exp\{(\Delta H_m/R)\,[(T_s - T_m)/T_s T_m]\}. \tag{1.8}$$

When N_2 is small, i.e. for low solubilities ($T_s \ll T_m$), this can be expanded to

$$N_2 = 1 + (\Delta H_m/R)\,(T_s - T_m)/T_s T_m. \tag{1.9}$$

As an example, the solubility at 25°C of a substance with a melting point of 150°C and a latent heat of melting of 2000 cal, will be

$$N_2 = 1 - [10^3(423 - 298)/(298 \times 423)] = 1 - 0.991$$
$$= 0.009 \text{ moles per } 0.991 \text{ moles of solvent.}$$

Examination of eqns. (1.6) and (1.9) shows that the solubilities will increase with increasing temperature.

Ideal solution behaviour, however, ceases to be displayed if (a) hydrogen bonding occurs between solute and solvent, (b) the solute molecules are much larger than the solvent molecules, or (c) the solute ionizes. The complexity of behaviour with non-ideal solutions is illustrated in Fig. 1.2, which shows the solubilities in water of phenol, triethylamine, nicotine, and ethyl ether in dependence on temperature.

At moderate temperatures, mixtures of phenol and water between about 10% phenol and 65% phenol separate into two phases, the

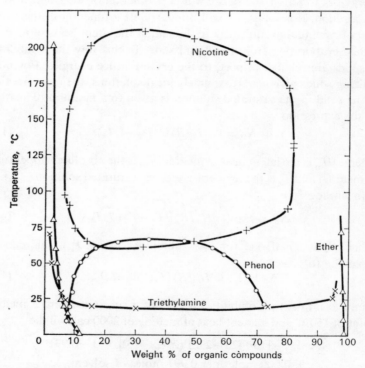

FIG. 1.2. Solubility curves for some organic compounds in water.

compositions of which are defined by the points at which the line of constant temperature cuts the solubility curve. With increasing temperature the solubilities of phenol in water and of water in phenol increase so that the compositions of the two phases approach each other, becoming identical at 65.8°C. At this temperature and above, phenol and water are miscible in all proportions. 65.8°C is the "lower consolute temperature" for phenol and water.

At moderate temperatures, mixtures of triethylamine and water between about 2% and 97% triethylamine separate into two phases. As the temperature *decreases*, the compositions of the two phases approach each other, slowly at first but very rapidly as the temperature falls to 20°C, and at 18.2°C the two phases become identical in composition: triethylamine and water are miscible in all proportions at 18.2°C and lower temperatures. 18.2°C is the "upper consolute temperature" for triethylamine and water.

With mixtures of nicotine and water between about 7.5% and about 80% and between temperatures of 61°C and 210°C, two phases are again formed: nicotine and water are miscible in all proportions below 61°C and above 210°C.

With less than 25% nicotine in water and temperatures between 61°C and 100°C, or with more than 30% nicotine in water and temperatures between 150°C and 210°C, the solubility of nicotine in water decreases with increasing temperature. With more than 30% nicotine in water and temperatures between 61°C and 100°C, or with less than 25% nicotine in water and temperatures between 150°C and 210°C, the solubility of nicotine in water increases with increasing temperature.

The curves for ethyl ether in water resemble the side branches of an even wider loop than the nicotine–water curve. Data for the lower part of the loop are unobtainable, however, because of the formation of ice, and for the upper part because the critical temperature of ether is reached first. In the curve to the left of the diagram, the solubility of ether in water is shown to decrease with increasing temperature: in the curve to the right of the diagram, the solubility of water in ether is shown to increase with increasing temperature. At 70°C the solution contains 0.08 mole of water per mole of ether.

It is shown in thermodynamics that

$$\partial \ln N_2/\partial T = \Delta H_{soln}/RT^2, \tag{1.10}$$

2*

where ln N_2 is the natural logarithm of the mole fraction of solute and ΔH_{soln} is the molar heat of solution. Thus if ΔH_{soln} is positive (i.e. heat is absorbed when the solute dissolves), the temperature coefficient of solubility is positive; the solubility increases with increasing temperature. If ΔH_{soln} is negative (i.e. heat is evolved when the solute dissolves), the temperature coefficient of solubility is negative. The former condition holds for ionizing tensides;[†] the latter condition holds for non-ionic tensides.

Solubility of Tensides

Non-ionic tensides, as will appear later, have the structure R.X. $(CH_2CH_2O)_nH$, where R is a hydrocarbon chain of at least 10 C atoms and XH is either —OH, —COOH, —SH, —NH_2, or —NH(R'). They thus contain a number of ether links between pairs of ethylene groups. Like ethyl ether–water systems, at high temperatures mixtures of non-ionic tenside and water tend to separate into two phases—a very dilute solution of tenside in water of the order of about 0.005%, and a solution of water in non-ionic tenside of the order of about 20% water or approximately 10 moles of water per mole of tenside. On such a diagram as those of Fig. 1.2 the turn into the nearly horizontal bottom branch of the curve passing through the upper consolute temperature would be extremely sharp and on the scale of the graph would be

---

† For the broad group of compounds variously classed as wetting, foaming, emulsifying, dispersing, and cleansing agents, no one name can be adequate. The terms surface active agent, surfactant, detergent, agent tension-actifs, and so on have been widely used but none has secured general adoption alone. At the 3rd International Congress on Surface Activity at Cologne an international committee on nomenclature officially adopted the German-proposed term "tenside". A technical–scientific journal with the title *Tenside* was launched shortly after and has achieved a secure place on the market. The term is short, not unduly cacophonic, and is adopted here although the seal of general international usage has not yet been given it. It may be objected that some surface active materials, especially some dispersing agents, have little effect on surface and interfacial tensions. They are effective because they form viscous or plastic barrier layers. The objection must, however, be dismissed as unduly pedantic.

scarcely separable from the left-hand vertical axis. The right-hand branch would also be steep at about the 80% ordinate. The reason for the extreme sharpness of the transition on the left-hand side of the curve is that when the concentration of the dilute solution phase has increased to a certain level, the structure of the solution changes from a sparse distribution of single molecules of non-ionic tenside in water to clumps of molecules in which the insoluble hydrocarbon portions collect together as a liquid drop, and the water-soluble ether groups and the terminal hydroxyl groups lie in the surrounding water. These clumps of molecules are called micelles. The micelles are very highly soluble in water. Owing to this structural feature of the solution, non-ionics either form a two-phase system or are miscible with water with a sharp temperature of transition between the two states.

The solubility curves for phenol, triethylamine, nicotine, and ethyl ether have portions in which the solubilities of the organic solutes increase and decrease respectively over the same temperature interval, depending on the mole fraction involved. For phenol and triethylamine in water, the rising and falling parts of the curve occur on opposite sides of the diagram: for nicotine and water, the rising and falling parts of the curve occur on opposite sides of the diagram for different temperature ranges. These facts demonstrate that to "explain" the changes in solubility as due to "changes in hydration" is to use an empty phrase. Such an "explanation" is no more valid in the case of non-ionic tensides than in the case of nicotine: unfortunately, it has a rather wide currency in the literature.

For ionizing tensides having the structure

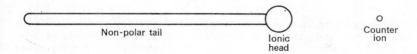

Non-polar tail   Ionic head   Counter ion

the energy of interaction between the ions and water is much larger than the energy of interaction between the ether dipoles of a non-ionic tenside and water. The ionic head of the molecule thus has a greater tendency to drag the non-polar tail into the water.[†] Within the temperature range 0–100°C (and, to an undetermined extent, beyond) ionizing tensides have a positive coefficient of solubility with increasing temperature. They are, like most uni-univalent salts, substantially completely ionized in water. At low temperatures they dissolve to a dilute solution of individual ions. As the temperature increases, the concentration in solution increases slowly until a characteristic temperature is reached at which the solubility begins to increase very steeply. As already explained in connection with the similar behaviour of non-ionic tensides, on lowering the temperature (because of their negative coefficient of solubility with rising temperature) the sudden change in the rate of increase of solubility is due to the formation of highly soluble micelles in which the paraffin tails are tucked away into the interior and the water-seeking ionic groups are distributed over the surface.

The phenomenon was investigated by Krafft many years before the present understanding of the micellar structure of the solutions was arrived at. He noticed that the temperature at which high solubility of the sodium salts of the higher fatty acids began was approximately equal to the melting points of the corresponding fatty acids. The correlation, though interesting, is fortuitous and is not observed with the potassium, ammonium, etc., salts. Because of his early investigations, the temperature at which the solubility of an ionic tenside reaches the "critical concentration for micelle formation" or "critical micelle concentration", the c.m.c., is often referred to as the "Krafft point".

The actual value of the c.m.c. depends strongly on the size of the non-polar tail of the tenside molecule but only to a small extent on the nature of the ionized "head group" or the associated uni-valent counter-ion. Thus in the straight-chain paraffin series of ionic tensides, the

---

[†] This type of structure was termed "amphipathic" in 1935 by Hartley (from Greek *amphi* = both, and *pathein* = be compatible with, suffer with). Winsor, in 1944, proposed to substitute the term "amphiphilic" (Greek *philos* = loving). The earlier term is to be preferred, not merely on ground of priority, but as having a less anthropomorphic connotation. It also avoids suggesting the antonym "amphiphobic" which would denote a physical impossibility. Further, the discussion above on the universal forces of attraction shows that all molecules are "philic" to every other molecule. "Amphiphilic" thus fails to distinguish one type of molecule from any other.

c.m.c. is, very approximately, about 0.01 M for the C_{12} compounds, 0.001 for the C_{14} compounds, and 0.0001 for the C_{16} compounds.

The solubility curves themselves vary not only with the size of the non-polar part of the molecule, but also with its structure, with the type of counter-ion, and with the crystal form with which the solution is in equilibrium. This is illustrated in Fig. 1.3.

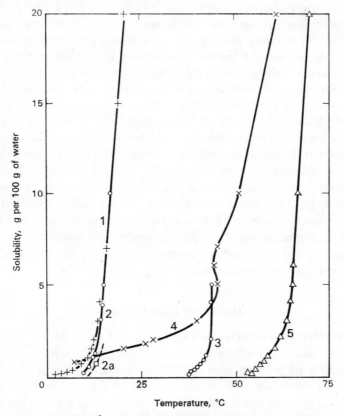

FIG. 1.3. Solubilities of some anionic tensides in water. 1, Sodium TPBS and ammonium n-dodecyl-benzene-p-sulphonate. 2, Sodium dodecyl sulphate. 2a, Sodium dodecyl sulphate from 2. 3, Sodium cetyl sulphate. 4,Ammonium TPBS.5,Sodium n-dodecyl-benzene-p-sulphonate.
TPBS = tetrapropylene benzene sulphonate.

Curves 2 and 2a relate to a highly pure sample of sodium dodecyl sulphate, the point 2a belonging to a different crystallographic form from that normally obtained by crystallization from aqueous alcohol. The lower solubility of this form compared with that of the "normal" form shows that it is the more stable variety.

Curves 1 and 4, for the sodium and ammonium salts respectively of tetrapropylene benzene sulphonate (TPBS) illustrate the striking difference that can be observed between salts of the same anionic tenside with not very dissimilar cations. The lowest point shown on curve 4 is already at a higher concentration level than the c.m.c. for the tenside. The slight bend in the curve showing two distinctly different solubilities at about 44°C is frequently observed with mixtures of tensides or impure products. TPBS comprises a large number of highly branched dodecyl isomers and some higher and lower homologues.

A difference between sodium and ammonium salts, in the opposite direction, is shown in comparison of curve 1 for ammonium n-dodecyl benzene sulphonate with curve 5 for sodium n-dodecyl benzene sulphonate, the ammonium salt becoming highly soluble at a much lower temperature than the sodium salt. The solubility curve of the highly branched sodium TPBS is indistinguishable on the scale of the graph from that of the straight-chain ammonium n-dodecyl benzene sulphonate and is in marked contrast with that of the straight-chain sodium n-dodecyl benzene sulphonate.

Solubility of High Polymers

The solubility characteristics of high polymers exhibit even more extreme features. Either the polymer dissolves to a viscous solution or the system of polymer plus solvent separates into two phases: an extremely dilute solution of polymer in solvent and a solution of solvent in polymer which takes the form of a swollen gel. The concepts of the "internal pressure" of a liquid, its "cohesive energy density", and the "solubility parameter" are particularly useful.

To separate the molecules of a liquid sufficiently to expand the molar volume by 1 cc at constant temperature requires the expenditure of a considerable amount of work, i.e. the system is raised to a higher potential energy level. The work done may be considered as being opposed by an "internal pressure" which is obviously related to the energy of cohesion (see the section below on surface tension, p. 29) and must be a function of the coefficients of thermal expansion at constant pressure, $(\partial V/\partial T)_P$, and of the coefficient of compressibility, $(\partial V/\partial P)_T$, at constant temperature. These coefficients are conveniently given the symbols α and β respectively.

Thermodynamics shows that

$$(\partial E/\partial V)_T = T(\partial P/\partial T)_V - P = \text{the internal pressure,} \quad (1.11)$$

where $(\partial E/\partial V)_T$ is the rate of increase of potential energy with increase in volume at constant temperature and $(\partial P/\partial T)_V$ is the rate of increase of pressure with increase in temperature at constant volume. Since the first term on the right-hand side, for normal liquids, ranges from 2000 to 8000 atm, the 1 atm of the second term P, the external pressure, may be neglected, hence

$$(\partial E/\partial V)_T = T(\partial P/\partial T)_V, \quad (1.12)$$

but

$$(\partial P/\partial T)_V = -(\partial V/\partial T)_P (\partial P/\partial V)_T = \alpha/\beta \quad (1.13)$$

(a standard relation in partial differentials of three variables), whence

$$(\partial E/\partial V)_T = T\alpha/\beta. \quad (1.14)$$

In words, the rate of increase of potential energy per mole with increase in volume at constant temperature—the internal pressure—is equal to the absolute temperature multiplied by the quotient of the coefficient of isothermal expansion at constant pressure and the coefficient of isothermal compressibility.

The internal pressure is also a function of the molar heat of evaporation per cc, i.e.

$$(\partial E/\partial V)_T = n \, \Delta H_V/V, \quad (1.15)$$

where V is the molar volume, ΔH_V is the molar heat of evaporation, and n is a numerical factor.

For many liquids n is close to unity, but for some it is very different: $\Delta H_V/V$ is therefore sometimes identical with the internal pressure, but it is not always so. On this account it is distinguished from the internal pressure by calling it the "cohesive energy density". The square root of the cohesive energy density has been found to afford a criterion for the mutual miscibility of two liquids. The symbol δ is used for $2\sqrt{(\Delta H_V/V)}$ in general. If δ_1 and δ_2 are written for this function for liquids 1 and 2, the critical consolute temperature of two liquids forming "regular" solutions (the liquids mix without heat change although the solution is not "ideal") is given by

$$T_c = (2N_1N_2V_1^2V_2^2)\,(\delta_1-\delta_2)^2/[R(N_1V_1+N_2V_2)^3]. \qquad (1.16)$$

When $(\delta_1-\delta_2)$ is less than the value required to satisfy eqn. (1.16), the two liquids are miscible.

If the mole fractions N_1 and N_2 are both put equal to 0.5 and the molar volumes V_1 and V_2 are put equal to each other, eqn. (1.16) simplifies to

$$T_c = (\delta_1-\delta_2)^2V/2R, \qquad (1.17)$$

where R is the gas constant, 1.987 cal/mole.

From eqn. (1.17) it is found that for T_c at 25°C (295°K), for liquids of molar volume 50 cc, $(\delta_1-\delta_2) = 4.0$, and for liquids of molar volume 100 cc, $(\delta_1-\delta_2) = 3.5$; liquids whose δ differences are less than these values will be miscible. Note that as the molar volume rises, the maximum permissible δ difference for miscibility falls. Because of the importance of the δ functions in solubility relationships, they have been named the "solubility parameters". Some values of the solubility parameters are listed in Table 1.1.

Co-solvency, Hydrotropy, and Solubilization

Within the general area of solvent phenomena it has proved convenient to distinguish three selected special cases: co-solvency, hydrotropy, and solubilization.

TABLE 1.1. SOME SELECTED VALUES OF THE SOLUBILITY PARAMETERS δ AT 25°C

Substance	δ	Substance	δ
Perfluoropentane	5.5	Polyvinyl acetate	9.4
Teflon	6.2	Polyvinyl chloride	9.53
2,2-Dimethylpropane	6.25	Carbon disulphide	10.0
n-Butane	6.7	Acetone	10.0
n-Pentane	7.05	Ethyl cellulose	10.3
n-Hexane	7.30	Acrylonitrile	10.5
Ethyl ether	7.40	Pyridine	10.7
Isoprene	7.45	Methyl cellulose	10.8
Polyethylene	7.9	Cellulose diacetate	10.9
n-Hexadecane	8.0	Epoxy resin	10.9
Perfluorobenzene	8.1	Methylene iodide	11.8
Natural rubber	8.35	Dimethyl formamide	12.1
Polybutadiene	8.45	Nitromethane	12.6
Butyl acetate	8.5	Ethanol	12.7
Buna S	8.55	Nylon 66	13.6
Polystyrene	8.56	Dimethyl sulphoxide	14.0
Carbon tetrachloride	8.6	Methanol	14.5
Benzene	9.15	Ammonia	16.3
Chloroform	9.3	Water	24.2
Styrene	9.3		

It is well known that mixtures of solvents may often be more effective than the individual solvents. Perhaps the best-known example is that of nitrocellulose (collodion) which is insoluble in either ether ($\delta = 7.4$) or ethanol ($\delta = 12$) alone, but is readily soluble to a viscous solution in a mixture of the two (δ about 10). The carboxylate soaps, and many other tensides, are more soluble in aqueous alcohol than in either water or alcohol. In 40% aqueous alcohol the colloidal character of ordinary soaps disappears, the micellar structure characteristic of the aqueous solution breaking down completely. The phenomenon is called co-solvency.

Many organic salts, tensides of low to moderate molecular weight and of high solubility, possess the property of enhancing the aqueous solubility of slightly soluble substances. This phenomenon has been extensively investigated by Neuberg who, in 1916, coined the term "hydrotrope" for these substances. Alkyl benzene sulphonates are particularly susceptible to the action of hydrotropes, of which group of substances sodium xylene sulphonate is an outstanding example. It is generally true to say that, as in the more strongly surface active tensides, the molecular structure must be that of distinctly separated polar and non-polar portions. For example, sodium benzoate and sodium *ortho* hydroxy benzoate are good hydrotropes, sodium *para* hydroxy benzoate is not. Aliphatic sulphates too short to form micelles, e.g. sodium hexyl sulphate or octyl sulphate are also excellent hydrotropes, the property being a maximum at about the eight-carbon atom chain length for various ionizing water-soluble compounds, and at about the four- to six-carbon chain length for non-ionizing compounds. The initial few per cent of hydrotrope often exerts a "salting out" effect, and frequently it is only in the presence of, say, 2–4% of a simple electrolyte that the change over to "salting in" or "hydrotropy" occurs. The hydrotropic effect of such substances as sodium xylene sulphonate on alkyl benzene sulphonates is enhanced in the presence of the water-insoluble tensides widely used as "lather improvers".

These properties are illustrated by the curves shown in Fig. 1.4, which show how the temperature at which commercial sodium TPBS dissolves to 10% changes with increasing hydrotrope concentration. The sample of sodium TPBS used was almost salt-free, the 10% solution containing only 0.1% of sodium chloride and 0.3% of sodium sulphate.

In the absence of other additives, the sodium TPBS dissolved to 10% at 21°C. Addition of 5% of a 95% pure, commercial lather improver, "coco"-diethanolamide (CDEA, often called "lauric" diethanolamide although "lauric" is strictly the trivial name for pure dodecanoic acid. Sometimes "lauroyl" will be found in the literature, by analogy with benzoyl), $RCO.N(CH_2CH_2OH)_2$, where RCO is the radical of coconut fatty acids, raised the solution temperature to about 40°C. Addition of 4% sodium sulphate raised the solution temperature to 70°C.

Sodium xylene sulphonate (NaXS), up to 15%, salts out the "pure" sodium TPBS, the undulatory shape of the curve suggesting a rather

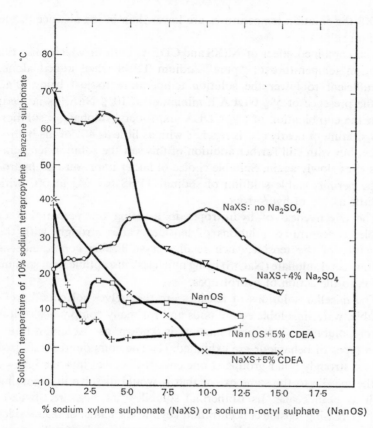

FIG. 1.4. Effect of hydrotropes on solubility. CDEA = "coco"-diethanol-amide.

nicely poised balance between the normal salting out effect of electrolytes and the salting in effect of hydrotropes. Sodium n-octyl sulphate also tends to both salt-out and salt-in, but in this instance the salting-in tendency is the stronger.

In the presence of sodium sulphate, sodium xylene sulphonate strongly reduces the temperature of solution. With more than 5%

NaXS the solution temperatures are lower than in the absence of the sodium sulphate.

The combined effect of NaXS and CDEA, both of which raise the solution temperature of "pure" sodium TPBS when added alone, is sufficient to lower the solution temperature to just below 0°C. In the presence of 5% CDEA a minimum of 10% NaXS is needed; with the combination of 5% CDEA and n-octyl sulphate, a solution temperature of nearly 0°C is reached with as little as 4% of the hydrotrope, but with still further addition of this salt the solution temperature rises slowly again. Suitable choice of lather improver and hydrotrope permits stable solution of sodium TPBS to 10% at 0°C with as little as 1% of hydrotrope.

The effectiveness of hydrotropes in assisting the preparation of stable, concentrated solutions of tensides varies strongly with the structure of the tenside, such readily crystallizable compounds as sodium cetyl sulphate (NaCtS) being much less susceptible than sodium TPBS to the action of hydrotropes.

The micellar solutions of tensides are themselves solvents for oil-soluble, water-insoluble compounds and for many compounds of low water solubility. This phenomenon is known as "solubilization". Two types of behaviour are exhibited. (1) The compound solubilized has no strongly polar groups at one end. It dissolves into the interior of the micelle to the same extent that it would dissolve in a paraffin such as hexadecane. Its isothermal solubility increases rectilinearly with increasing tenside concentration beyond the c.m.c. of the tenside: azobenzene behaves in this way. (2) The compound solubilized has a strongly polar group at one end, as in the dyestuff yellow AB, the polar group of which remains in the water.

Compounds with this type of structure tend to show an increasing solubility per mole of tenside with increasing concentration of tenside.

Adsorption

The interplay of the forces between molecules which govern solubility, as was shown earlier, also leads to the phenomenon of *adsorption*. Wherever the aqueous solution meets a boundary surface the molecules of the tenside may be orientated so that the tails are extruded from the water while the water-solubilizing groups remain immersed in the water. Since location of a molecule in such a position is energetically favoured compared with location of the molecule within the aqueous medium, the tenside molecules will tend to accumulate in the surface in greater concentration than in the bulk solution. This greater surface concentration compared with bulk concentration is called adsorption. As the surface concentration builds up, the rate at which tenside molecules return from the surface to the bulk by the ordinary process of diffusion will also increase, equilibrium being reached when the numbers of tenside molecules entering and leaving the surface in a given time interval are equal.

At the air–water interface, the tenside molecules arriving first will tend to lie with most of their hydrocarbon tails lying flat along the surface, lifting up from the surface into the air at statistically extremely infrequent and for very brief intervals. As they increase in number, the tails—which are, of course, still undergoing vigorous molecular motion—tend to become more and more entangled. However, the surface film of ionized, water-soluble tenside does not become close-packed because the mutual repulsion of the charged head-groups prevents so dense a surface population being formed. The mutually entangled tails are to be thought of as in the state of a two-dimensional vapour of high concentration, or as a two-dimensional liquid. The conventional diagram showing the tenside tails rising from the water surface like fence posts, all carefully aligned, is easy to draw but entirely fictitious.

Addition of very poorly soluble, polar, non-ionizing tensides to aqueous solutions of ionic tensides permits a considerable increase in the total surface population of tenside molecules, the polar non-ionic molecules inserting themselves into the surface between the ionized molecules. The "concentrated vapour" type of surface film becomes fully liquid. The familiar carboxylate soaps provide such a mixed surface composition by virtue of their partial hydrolysis to free fatty acid. Although the proportion of hydrolytic fatty acid in the soap solution may be very small, the surface film usually consists of about equimolar portions of soap and fatty acid. (Note that at a pH at 25°C of 10.5, due to hydrolysis, a 5% solution of a toilet soap of average molecular weight about 264 is only about 0.17% hydrolysed.) The non-ionic tensides which can perform this function form the large group of "lather improvers" and "emulsifying aids".

At an oil–solution interface, tenside molecules will also tend to crowd into the surface. In this instance, however, the tails are not obliged to remain flat at the interface but may penetrate into the oil phase, the only requirement being that the water-solubilizing groups remain in the water. Apart, therefore, from being anchored to the interface at one end, the tails are liquid paraffinic molecules within the liquid oil phase, bending, twisting, and moving about freely. The mutual repulsion of the charges of ionized tensides again tends to limit the density of population of the interface and, again, this can be overcome by the addition of suitably chosen polar, non-ionizing tensides. The ionizing tensides will frequently be somewhat oil-soluble, and the polar non-ionic tensides will generally be highly oil-soluble. There are, therefore, complex partition equilibria of tensides between the two bulk phases and the interface.

Water will also squeeze out the non-polar tails of the tenside molecules on to the majority of solid–solution interfaces. The exceptional solid surfaces are those which strongly attract water themselves. Clean cotton, for example, does not adsorb alkali metal salts of anionic tensides, but will take up small amounts of alkaline earth or heavy metal salts. This is due to the carboxyl groups present on cellulose formed by oxidation of some of the primary hydroxyl groups, the number of carboxyl groups depending on the age and previous treatment of the cotton.[†]

The divalent metal acts as a bridge between the carboxyl group on the cotton and the tenside anion. The calcium present in water of 24°H (French) or residing on the cotton after previous treatment in hard water is sufficient to give an apparent adsorption of sodium salts of tensides, in the latter case even when the tenside is used in distilled water.

Since cotton is negatively charged, cationic tensides are strongly adsorbed on it. The same is true of glassware. In such cases the tenside is first adsorbed "heads down" with formation of a hydrophobic surface of non-polar tails. A second layer of molecules is readily adsorbed on the first "tails down", leaving a hydrophilic surface of head groups exposed to the water. This is readily demonstrated by washing glassware with a solution of cetyl trimethyl ammonium bromide. The glassware is washed sparklingly clean and the residual solution drains to a thin continuous film when the bulk has been poured away. The first two or three rinse waters also drain to thin films but, finally, the outer layer of adsorbed molecules having been rinsed off, the rinse water drains leaving large hanging drops on the glass just as if the glass were greasy.

Ionic tensides interact strongly with proteins. Lundgren found that feather keratin could be dissolved in concentrated aqueous anionic tensides and the resulting solution could be re-precipitated in continuous fibre form by extrusion through spinnerets into a concentrated sodium sulphate bath. The tenside could be extracted from the fibre

---

[†] Even the most carefully prepared cotton contains carboxyl groups produced by oxidation of the primary hydroxyl groups not only in the course of the commercial manufacture of the fabric from the raw cotton, but during the growth of the fibre in the plant as well. In paper technology much attention has been given to the determination of the number of carboxyl groups per 100 glucose rings. Removal of heavy metals may be achieved by treatment with polyphosphate or other "sequestering agent" followed by thorough rinsing in de-ionized water. The writer has found that not less than twenty rinses may be necessary. The carboxyl groups may then be titrated with caustic soda and phenolphthalein **in the presence of about 0.1 N sodium chloride.**

with 70% acetone–water. Pankhurst and his colleagues carried out an exhaustive investigation on the interaction of a soluble protein (gelatin) with sodium dodecyl sulphate. In alkaline solution the complex remained in solution: in acidic media the complex was precipitated. Enzymes, in which a protein is an important part of the molecule, may be inactivated by tensides.

It has long been known that some soaps are more irritant to the skin than others. The same is true of the long-chain sulphate semiesters and the alkyl-aryl sulphonates, which form the bulk of commercial tensides. It is reasonable to postulate that an important property of tensides related to their irritancy to the skin is their capacity to interact with the skin proteins. This interaction is readily measured by measuring the adsorption. Such adsorption will modify, perhaps profoundly, the properties of the epidermis, e.g. its water-holding capacity, its permeability to water and to other molecules, and the accessibility to reaction of chemical groups within the keratin molecule. Interference with enzyme systems and with the RNA/DNA[†] components of the living cells will also promote skin reactions.

The capacity of the skin to resist the action of tensides will depend on the structure of the tenside molecule, its concentration, the time of contact, the concentration of additional electrolytes, and the temperature. These features will all be reflected in the adsorption curves of the tensides, particularly by the kinetics of the adsorption.

Figure 1.5 shows a selection from a number of adsorption curves, for tensides on prepared hide powder.[‡] The data fit the equation

$$A_t = A_\infty\{1 - \exp[-\sqrt{(kt)}]\}, \tag{1.18}$$

where A_t is the adsorption at time t in minutes, A_∞ is the equilibrium adsorption, and k is the rate constant, which is consistent with diffusion being the rate-determining factor.

Comparison of curves 1, 4, and 8 shows the effect of changing the structure of the tenside. NaCtS is adsorbed much more slowly than

---

[†] Ribonucleic acid/deoxyribonucleic acid.

[‡] Made to the specification of the Society of Leather Trade Chemists as a standardized product: sold by Baird & Tatlock Ltd.

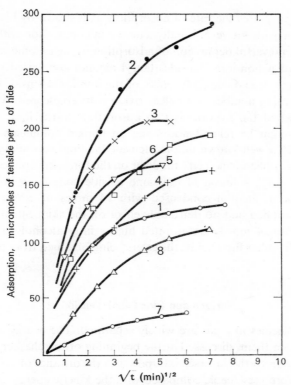

FIG. 1.5. Adsorption of tensides on hide powder. 1, Sodium TPBS: 0.005 M; 32°C; pH 7. 2, Sodium TPBS: 0.010 M; 32°C; pH 7. 3, Sodium TPBS: 0.00515 M. Na$_2$SO$_4$: 0.010 M; 20°C; pH 7. 4, NaLS: 0.005 M; 32°C; pH 7. 5, NaLS: 0.005 M; 45°C; pH 7. 6, NaLS: 0·005 M. Na$_2$SO$_4$: 0.0025 M; 32°C; pH 7. 7, NaLS: 0.005 M. Lauryl alcohol-9-3-ethylene oxide: 0.005 M; 45°C; pH 7. 8, NaCtS: 0.005 M; 32°C; pH 7.

NaLS = sodium lauryl sulphate.

NaLS and is known to be much milder in its effect on the skin. Sodium TPBS is adsorbed more rapidly than NaLS but tends to a much lower equilibrium adsorption: the two curves cross after about 5 minutes. Comparison of curves 1 and 2 shows the effect of doubling the concentration of tenside and comparison of curves 1 with 3 and 4 with 6 shows the effect of adding sodium sulphate. Increase of temperature

3*

markedly increases the rate of adsorption, as is shown by comparison of curve 4 with curve 5. Finally, curve 7 in comparison with curve 5 shows the powerful depression of adsorption of the anionic tenside on addition of a non-ionic tenside (lauryl alcohol condensed with 19.3 moles of ethylene oxide). The effect of the non-ionic tenside increases with increasing number of ethylene oxide units condensed. An experimental test of the hypothesis that adsorption and irritant effect on the skin might be related was made on human volunteers. An 8% solution of a well-known anionic tenside (sodium secondary alcohol sulphates) applied in a "cell" for 24 hr on the forearm produced severe lesions. An 8% solution of the same tenside containing in addition 8% of sperm alcohol condensed with 24 moles of ethylene oxide similarly applied had no detectable effect on the skin. Sperm alcohol is a mixture of mostly unsaturated higher fatty alcohols from their esters with higher fatty acids, the esters being obtained from the sperm whale.

Surface and Interfacial Tension

The molecules in a gas are widely separated. The energy of attraction between them, discussed at the beginning of the chapter, falls off in proportion to the inverse sixth power of the distance of separation: it is therefore very feeble compared with the kinetic energy of motion of the gas molecules and the gas has no cohesion. The molecules in solids and liquids are in "contact". In liquids the mean distance of separation between neighbouring molecules is usually greater than in the corresponding solid, but this is not universally true, the expansion of water on freezing being well known. Some metallic alloys, notably type metal, have the same property. In these condensed phases the energy of attraction is greater than the kinetic energy of thermal motion and a condensed phase differs from a gas in possessing the property of cohesion. Individual molecules are in a state of vibration about a mean position of equilibrium; in a liquid individuals change places readily; in a solid they do so very seldom although the phenomenon of solid diffusion shows that occasional interchange of position of molecules can and does occur even in solids. However, whereas diffusion in liquids is experimentally detectable in a time scale of a few seconds, in solids the time scale becomes many months or even years.

Notwithstanding the few exceptions at the melting point itself, the general picture is that, as the temperature of a solid is raised, the energy of thermal vibration increases and the mean distance of separation gradually increases. The frequency with which molecules interchange their positions increases, and finally the structural pattern suddenly breaks down and all the molecules become mobile—the solid has melted and become a liquid.

Molecules deep within the body of a liquid are subjected to attractive forces on all sides. Those close to the surface are subjected to the sum of the attractions of all the molecules below, to the side, and above them, but the number of molecules above is very small for molecules close to the surface and nil for those actually in the surface. The molecules in the surface are in a position of higher potential energy than those in the interior, and the surface will therefore tend to contract to a minimum.

Because the molecules in the surface of the liquid are in a state of higher potential energy than the molecules in the interior, work must be done to bring a molecule isothermally from the interior into the surface, and the energy involved is recovered when the molecule returns to the interior. This isothermal work, or free energy per molecule, multiplied by the number of molecules in 1 cm² of surface, is the free energy of formation of 1 cm² of surface. It is thus the *excess* energy which surface molecules possess compared with that of an equal number of molecules in the interior. Owing to a small density difference between surface layer and interior, this is not quite the same thing as the molecules in 1 cm² of surface compared with 1 cm² of a plane in the interior.

If a liquid surface is physically restrained, the tendency of the system towards a position of equilibrium will be manifested as a tension in the surface of the liquid. This "surface tension" is mathematically equivalent to the excess Helmholtz free energy of formation of 1 cm²

of surface;[†] the work done in isothermally extending the surface by
a cm along a line of b cm length against the surface tension γ dynes/cm
is $ab\gamma$ ergs: the free energy of formation of an area of surface ab cm²
is $ab\gamma$ ergs, where γ is now the free energy of formation in ergs/cm².
The discussion which has persisted in the literature for many years
as to whether the "tension" or the "energy" is real and the other
derived is a purely philosophical question. Force concepts and energy
concepts are two different languages with which the phenomena may
be described.

Molecules in the surface of a solid are also in a state of high potential energy relative to those in the interior. Further, apart from the
cube faces of cubic crystals, the potential energy will be different on
the different crystal faces. The specific surface free energy is still the
free energy of formation per square centimetre, but, in the case of
solids, this is not equal to the work of extension since a solid cannot be
extended without setting up elastic strains throughout the solid, and
additional work must be done against the elastic restoring forces of
the solid to set up these strains.

When two condensed phases are brought into molecular contact,
the molecules in the two surface layers will be subject to forces across
the boundary and hence will be more nearly in the state of interior
molecules. However, unless the molecules in the two phases are
identical, the surface layers will, in general, not be in quite the same
state as molecules in the respective interior phases. There will, therefore, be residual excess free energy in each of these surface layers.
The total excess free energy γ_{12} of the interface between the two phases
will be the sum of the two. Thus

$$\gamma_{12} = \gamma_1' + \gamma_2', \tag{1.19}$$

† This is the difference between the free energy of the real system consisting of
liquid in equilibrium with its vapour, including the interface between the two phases,
and the free energy of a hypothetical system of the liquid and its vapour, at the
equilibrium pressure, without an interface between them. If a narrow cylinder
containing a small volume of liquid is laid isothermally on its side, the area of liquid
is increased at constant volume of the system, hence it is the Helmholtz free energy
change which is the free energy of formation of fresh surface.

where γ_1' and γ_2' are the residual excess surface free energies of each phase at the interface, not the excess free energies at the condensed phase–air boundary surface.

Surface Tension of Solutions

The adsorption of solutes, already briefly referred to, leads to changes in the surface tension compared with that of the pure solvent. The change in surface tension with increasing concentration of solute is given by the famous Gibbs adsorption theorem:

$$dy = -S_s dT - \Gamma_1 d\mu_1 - \ldots - \Gamma_i d\mu_i, \qquad (1.20)$$

where dy is the differential of the surface tension and S_s is the excess entropy of the surface, i.e. is the difference between the entropy of the real system and that of a hypothetical system in which the concentration of each species in each phase remains constant up to the physical boundary of separation between the liquid and vapour phases. Excess free energy and excess enthalpy of the system are similarly defined and are frequently referred to as the surface free energy and the surface free enthalpy respectively.

$\Gamma_1, \ldots, \Gamma_i$ are the surface excesses of the adsorbed components, 1, 2, 3, $\ldots$, i, and $d\mu_1, \ldots, d\mu_i$ are the differentials of the chemical potentials of the components 1, 2, 3, $\ldots$, i.

At constant temperature, the entropy term vanishes and, for only two components, the equation becomes simply

$$dy = -\Gamma_1 d\mu_1 - \Gamma_2 d\mu_2. \qquad (1.21)$$

It is possible to choose the mathematical diving plane which defines the surface so that the surface excess of the solvent is zero,[†]

[†] An extended discussion would show that a number of conventions may be used to define the surface excess. The convention used here compares the concentration of solute in a volume which includes the surface with its concentration in a volume in the interior which contains the same number of moles of solvent. The different conventions are analogous to the mole ratio, mole fraction, volume ratio, and volume fraction conventions which may be used to describe bulk concentrations.

hence

$$dy = -\Gamma_2^1 \, d\mu_2$$
$$= -RT\Gamma_2^1 \, d(\ln f_2 N_2)$$

or
$$\Gamma_2^1 = -(f_2 N_2 / RT) \, [dy/d(f_2 N_2)], \tag{1.22}$$

where Γ_2^1 is to be read "the surface excess of the solute on convention 1", f_2 is the activity coefficient of the solute, N_2 is its mole fraction, ln is the natural logarithm, R is the gas constant, and T is the absolute temperature (°K).

For dilute, ideal solutions, the activity coefficient is equal to unity, hence the mole fraction is proportional to the concentration c_2, whence

$$\Gamma_2^1 = -(c_2/RT) \, (dy/dc_2) = -dy/(RTd \ln c_2). \tag{1.23}$$

Thus if the surface tension falls with increasing concentration, adsorption of the solute is positive, whereas if the surface tension rises with increasing concentration, adsorption of the solute is negative. The latter occurs with solutions of many simple inorganic salts, the possible range of increase of surface tension being limited to only a few dynes per centimetre.

Positive adsorption occurs very commonly with aqueous solutions of organic substances, and is particularly marked with solutions of tensides. In such solutions, therefore, the surface tension is much lower than that of pure water.

On a molecular-kinetic view, the adsorbed molecules, being in a state of two-dimensional translation, vibration, and rotation, exert an outward pressure—the spreading pressure ∂. The surface tension of the liquid is reduced by the spreading pressure. The two-dimensional analogy with bulk matter may be pressed further. The adsorbed layer may form islands of a condensed surface phase in equilibrium with mobile surface molecules at a constant spreading pressure or surface vapour pressure of the surface condensed phase. If this occurs at a bulk concentration below the saturation limit, further increase in the bulk concentration may lead to an increase in the amount of surface condensed phase without changing the spreading pressure and hence

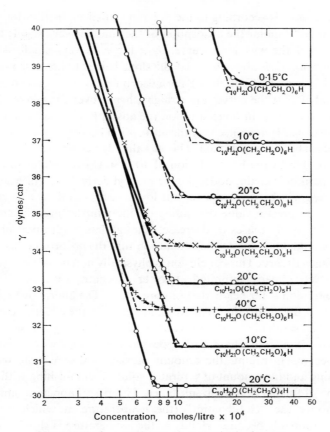

FIG. 1.6. Surface tension curves: non-ionics in water.

without changing the surface tension. It follows that the increase in bulk concentration cannot in such circumstances be accompanied by an increase in the bulk activity of the solute.

Figure 1.6 (data due to Hudson) shows the course of the γ vs. log c curve, over a wide concentration range, for pure dodecyl tetra-, penta-, and hexaethylene oxide condensates (commercial non-ionic tensides are mixtures in which the distribution of numbers of condensed ethylene oxide units is according to the Poisson statistical distribution law). At about 10^{-6} molar the adsorption of the solute is very small; hence the slope of the γ vs. log c curve, $d\gamma/d$ log c, is very small. As the concentration rises above 4×10^{-6} the slope beings to increase rapidly, indicating a rapid build-up of the adsorption layer. Between about 10^{-5} and 10^{-3} M the curves are straight; hence over the whole of this hundredfold range in concentration the adsorption remains constant. It is a reasonable inference that the adsorption becomes constant when the surface becomes saturated. This has already occurred at the minute concentration in the bulk solution of 10^{-5} M. (With solutes having a longer paraffinic chain, dodecyl or tetradecyl derivatives, for example, the curves will be displaced to still lower concentrations.) At about 10^{-3} M, the precise value depending on the temperature, the surface tension suddenly ceases to decrease. Thoughtless application of eqn. (1.23) would lead one to the conclusion that the adsorption had suddenly fallen to zero. This conclusion is physically untenable since (a) the surface was already saturated at lower concentrations, and (b) the surface tension is far below that of pure water. The alternative suggestion is that the adsorption remains unaltered, but the effective concentration has suddenly become invariant with increasing concentration of solute. If a fresh phase had appeared the system would, indeed, have become invariant, the amount of, say, wet non-ionic tenside increasing and the amount of saturated solution diminishing with each addition of further non-ionic substance. This does happen at temperatures above the "cloud point" (the temperature at which a 0.5% solution suddenly becomes cloudy as the temperature is slowly raised) of the solute. In the present instance, however, the change is not quite the separation of a new phase, but the formation of micelles, which is very nearly equivalent to the formation of a new phase.

The interpretation of similar curves for ionizing tensides is complicated by the ionization. If the medium is not pure water but a salt solution of constant concentration, the effect of the ionization is swamped out and eqn. (1.23) still applies. In the absence of the salt, the organic ion and the small inorganic counter-ion, necessarily always at the same equivalent concentration in any given solution, are two components—not one. Analysis of the problem by Pethica has shown that, in this case, the equation becomes

$$\Gamma_2^1 = -2d\gamma/(RTd \ln c_2) \qquad (1.24)$$

(within the range of very dilute solutions where the activity coefficient is unity). At high concentrations eqn. (1.23) again applies.

Detailed discussion of the thermodynamics of these systems has recently been published by van den Tempel and by Lucassen–Reynders.

A typical curve for an ionizing tenside is shown in Fig. 1.7. Van Voorst Vader points out the conclusion (given above in the discussion

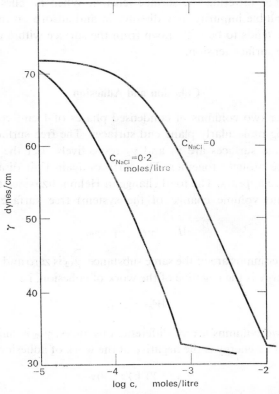

FIG. 1.7. Surface tension curves for sodium dodecyl sulphate.

of Fig. 1.6) to be drawn from the rectilinear portion of the graph for concentrations less than the concentration for micelle formation.

Much of the surface tension work reported in the earlier literature was vitiated by the unrealized effect of small amounts of such impurities as lauryl[†] alcohol in sodium lauryl sulphate. As little as 0.5% of such impurity shifts the lower part of the curve somewhat to the left and produces a noticeable minimum in the γ vs. log c curve in the neighbourhood of the c.m.c., the depth of the minimum increasing with increasing amount of impurity. At tenside concentrations less than the c.m.c. the insoluble impurity is strongly adsorbed at the surface, thus lowering the surface tension. When micelles form, the water-insoluble impurity may dissolve in and adsorb on the micelles, and hence tends to be withdrawn from the surface with a consequent rise in the surface tension.

Cohesion and Adhesion

Consider two columns of condensed phases of 1 cm² cross-section and having molecularly plane end surfaces. The free surface energies of the plane surfaces are γ_1 and γ_2 respectively. Let the two plane surfaces be brought together. The surfaces against air disappear and an interface appears. The total change in Helmholtz (since the process involves no volume change of the system) free surface energy is therefore

$$\varDelta F = \gamma_{12} - \gamma_1 - \gamma_2. \tag{1.25}$$

If the two columns are of the same substance, γ_{12} is zero and the change in free energy is the negative of the work of cohesion, i.e.

$$W_c = 2\gamma_1. \tag{1.26}$$

If the two columns are of different substances, γ_{12} is finite and the change in free energy is the negative of the work of adhesion, i.e.

$$W_a = \gamma_1 + \gamma_2 - \gamma_{12}. \tag{1.27}$$

[†] The trivial name for dodecanol.

There is no known example in which the interfacial excess free energy is greater than the sum of the surface excess free energies of the two substances. It is evident from the discussion of intermolecular energies above that it cannot be expected that such an example will ever be found. Thus *between any two condensed phases in molecular contact, there is always adhesion*, as, indeed, was already evident in the discussion leading to eqn. (1.3).

Table 1.2 lists some values for the work of adhesion of various organic substances to water in comparison with the work of cohesion.

TABLE 1.2. WORKS OF COHESION OF SOME
ORGANIC LIQUIDS (W_c) AND THEIR WORKS OF
ADHESION TO WATER (W_a) (AT 25°C)

Organic liquid	W_c (ergs/cm²)	W_a (ergs/cm²)
Paraffins	37–45	36–48
Aromatic hydrocarbons	60	63–67
Alkyl halides	50	66–81
Esters	50	73–78
Ketones	50	85–90
Nitriles	55	85–90
Primary alcohols	45–50	92–97
Fatty acids	51–57	90–100
Water	144	

Note that the works of cohesion of the different classes of organic liquid are closely similar and only little greater than that of the paraffins. The surfaces must therefore also be very similar. It requires more work to bring the polar part of the molecule into the surface than to bring the non-polar part. The great majority of the molecules in the surface are therefore, at any one instant, orientated with their paraffinic ends towards the air and their polar ends buried beneath the surface. In spite of the intense thermal agitation of the molecules, only a very few are momentarily rotated so as to expose their polar ends to the air. On the other hand, exposure of their polar ends to the water, at the organic liquid–water interface, is favoured by the interaction between the dipoles of the water and of the organic substance. This tends to reduce the interfacial tension γ_{12} and hence to increase the work of adhesion.

Equation (1.27) applies to liquid–liquid, liquid–solid, and to solid–solid interfaces provided only that the surfaces are in true molecular contact. This is easy to realize for liquid–liquid interfaces, fairly easy to realize for liquid–solid interfaces, and often very difficult to realize for solid–solid interfaces.

Spreading and Wetting

Consider a volume of liquid 1 placed on the surface of a large volume of liquid 2, with which it is immiscible, to form a circular drop of not less than 20 cm radius. (The necessity for this restriction to large drops is shown in the section below under linear tension.) As the drop of liquid 1 spreads, its area increases: the area of liquid 2 diminishes by the same amount, and an equal area of interface is formed. At some stage the vertical cross-section will have the form shown in Fig. 1.8a. The periphery is subjected to an outward force given by

$$\mathcal{S} = \gamma_2 - \gamma_1 \cos \alpha - \gamma_{12} \cos \beta. \qquad (1.28)$$

When the angles α and β become very small, $\cos \alpha$ and $\cos \beta$ tend to unity and $\mathcal{S}$ tends to

$$\gamma_2 - (\gamma_1 + \gamma_{12}) \qquad (1.29)$$

as α and β tend to zero. Thus the condition for continuous spreading of the liquid drop on the second liquid is that the sum of its surface tension and the interfacial tension is less than the surface tension of the second liquid. (A reversed $\mathcal{S}$ is used, following Harkins, to avoid confusion with S for entropy and for surface.) $\mathcal{S}$ is called the "spreading pressure". Comparison of the expression (1.29) with eqn. (1.27) shows that

$$\mathcal{S} = W_a - 2\gamma_1 = W_a - W_c \quad (\alpha = \beta \to 0), \qquad (1.30)$$

and therefore that an alternative statement of the criterion for initial

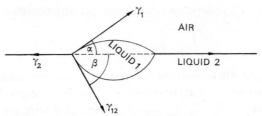

(a) Liquid lens on an immiscible liquid

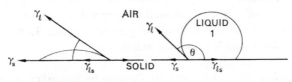

(b) Liquid drop on solid

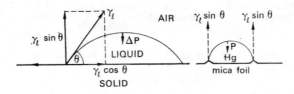

(c) Vertical component of liquid surface tension

Fig. 1.8. Liquid lens and drops and vertical component of surface tension.

continuous spreading is that the work of adhesion should be greater than the work of cohesion of the spreading phase.

It sometimes happens, particularly with impure oils on water, that the oil-drop spreads initially in accordance with eqn. (1.28) with $\alpha = \beta = 0$, but later the edge stops advancing and retracts. This may be due to "surface solution" of a component from the oil-drop spreading out ahead of the drop boundary until it has formed a complete monolayer and reduced γ_2 to γ_2'. γ_2' is no longer greater than $\gamma_1 + \gamma_{12}$ and the drop retracts and forms finite angles α and β at the boundary line until $\mathcal{S}'$, the final spreading coefficient, has become zero.

Equation (1.29) also shows that if the spreading coefficient be exactly zero,

$$\gamma_{12} = \gamma_1 - \gamma_2, \tag{1.31}$$

so that yet another criterion for spontaneous spreading is that the interfacial tension is equal to or greater than the difference between the surface tensions of the two liquid phases. Antonov postulated that eqn. (1.31), with the = sign only, holds as a general rule when the phases are mutually saturated. However, since both positive and negative values of $\mathcal{S}$ and $\mathcal{S}'$, as well as zero values, are known, the so-called "Antonov's rule" cannot be generally valid. Table 1.3 gives some examples where it is grossly untrue.

TABLE 1.3. SURFACE AND INTERFACIAL TENSIONS OF MUTUALLY SATURATED PAIRS OF LIQUIDS AT 25°C

Mutually saturated liquids	γ_1'/γ_2' (dynes/cm)	$\Delta\gamma'$ calc. (dynes/cm)	γ_{12}' obs. (dynes/cm)	Difference	Difference as % of calculated
Benzene–water	28.82/62.36	33.54	35.03	1.49	4.45
Carbon bisulphide– water	31.81/70.49	38.68	48.63	9.95	25.9
Methylene iodide– water	50.52/72.20	21.68	45.87	24.19	111.5
Heptanol–water	26.48/28.53	2.05	7.95	5.90	288
Isopentanol–water	23.56/25.92	2.36	5.00	2.64	112

Polar–polar pairs of phases show the greater deviations, less polar pairs show smaller deviations, polar–non-polar pairs still smaller deviations, and non-polar–non-polar pairs, if not miscible in all proportions, show little, if any, deviation.

In recent years Fowkes has proposed that surface tensions may be separated into contributions from the "dispersion" forces and the "polar" forces of molecular interaction. For non-polar molecules only the dispersion forces are operative and even for many polarizable and polar molecules they constitute a major part of the total intermolecular forces. It has been further postulated that Antonov's rule may hold generally if restricted to the dispersion force contribution to the surface and interfacial tensions. This may be a reasonable postulate.

On solids, a sessile drop takes the form shown in Fig. 1.8b. Resolution of forces in equilibrium at P, as shown in Fig. 1.8c, leads to the equation

$$\gamma_s = \gamma_{ls} + \gamma_l \cos \theta. \tag{1.32}$$

In the absence of equilibrium, by analogy with eqn. (1.28),

$$\mathcal{S} = \gamma_s - \gamma_l \cos \theta - \gamma_{ls}, \tag{1.33}$$

where $\mathcal{S}$ is the spreading coefficient of the liquid on the solid, γ_s is the surface tension of the solid, and γ_l is the surface tension of the liquid.

Equation (1.27) applied to the liquid–solid case, takes the form

$$W_a = \gamma_s + \gamma_l - \gamma_{ls},$$

and combination of this with eqn. (1.32) gives

$$W_a = \gamma_l (1 + \cos \theta) \tag{1.34}$$

showing that the work of adhesion of liquid to solid can be obtained from a measurement of the surface tension of the liquid and the angle of contact of the liquid on the solid. Further, if the angle of contact is just zero, the work of adhesion of the liquid to the solid is equal to twice the surface tension of the liquid, i.e. to its work of cohesion.

Hohn and Lange suggested in 1935 that Antonov's rule might usefully be extended to include the liquid–solid interfaces. This suggestion was taken up in 1937 by Powney and Frost. Writing eqn. (1.31) in the form $\gamma_{ls} = \gamma_l - \gamma_s$ and substituting in eqn. (1.32) leads to

$$\left. \begin{aligned} \cos \theta &= (2\gamma_s/\gamma_l) - 1 \\ \gamma_s &= \gamma_l (1 + \cos \theta)/2 \\ \gamma_{ls} &= \gamma_l (\cos \theta - 1)/2 \end{aligned} \right\} \tag{1.35}$$

thus leading to a value for the excess free surface energy of the solid, and hence also for a value for the solid–liquid interfacial tension γ_{ls}—quantities otherwise impossible to determine experimentally.

TABLE 1.4. CALCULATION OF THE SURFACE
TENSION OF PARAFFIN WAX

% sodium oleate $\times 10^3$	$\theta°C$	$\cos \theta$	γ_l	γ_s from eqn. (1.35)	γ_{ls}
3	68	0.375	36.2	24.9	11.3
4	65	0.423	35.3	25.1	10.2
5	63	0.454	34.4	25.0	9.4
6	60	0.500	33.6	25.2	8.4
7	55	0.574	32.8	25.8	7.0
8	50	0.643	32.0	26.3	5.7
9	48	0.669	31.4	26.2	5.2
10	46	0.695	30.8	26.1	4.7
11	43	0.731	30.3	26.2	4.1
12	40	0.766	29.9	26.4	3.5
20	34	0.829	27.2	24.9	2.3
40	25	0.906	25.7	24.5	1.2
50	0	1.00	25.7	25.7	0
	Mean value			25.5	

This was applied to calculate the surface tension of solid paraffin wax from the surface tensions of dilute aqueous solutions of sodium oleate and their contact angles on the wax. Table 1.4 shows the results obtained.

Zisman and his co-workers have carried out extensive and painstaking work on the contact angles between numerous liquids and carefully prepared surfaces of solids of low surface energy. They found that a plot of γ_l vs. $\cos \theta$ gave a substantially straight line for the liquids of a homologous series, although liquids of different chemical types gave values which lay within a band rather than on a line, and the band was curved for low values of $\cos \theta$, high values of γ_l. Figure 1.9 shows Powney and Frost's data of Table 1.4 on a Zisman plot. Apart from two clearly aberrant points, the points lie on a good straight line fitted by the equation

$$\cos \theta = 2.542 - 0.060\gamma_1$$

and extrapolating to $\cos \theta = 1$ ($\theta = 0$) at $\gamma_l = 25.8$.

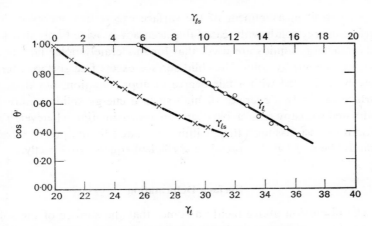

FIG. 1.9. Zisman plot of data of Powney and Frost.

Any liquid having a surface tension of 25.8 dynes/cm or less will spread spontaneously on paraffin wax. $\gamma_l = 25.8$ is the limiting surface tension for spreading, or, in Zisman's terminology, the critical surface tension γ_c.

It is obvious that if $\cos \theta = 1$ is put into eqn. (1.35),

$$\gamma_s = \gamma_l = \gamma_c \quad (\cos \theta = 1). \tag{1.36}$$

It is further obvious from eqn. (1.32) that, at this point on the graph

$$\gamma_{ls} = 0 \quad (\cos \theta = 1). \tag{1.37}$$

It would appear that in this procedure there is a means of arriving at values for the solid–air and solid–liquid superficial tensions, but it is essential to be quite clear as to the limitations involved. The deduction depends on the assumed validity of Antonov's rule, and it has already been shown that this rule is not of general validity. However, if Fowkes's postulates that for low surface energy substances (non-polar substances in general) only the dispersion forces are of significance for their surface tensions, and that under these conditions Antonov's rule is experimentally valid, then it may be concluded that eqns. (1.35) and

4*

(1.36) permit measurement of the surface excess free energy of low surface energy solids with reasonable accuracy provided only that low surface energy liquids are used, that for moderately polar liquids on low surface energy solids the solid surface excess free surface energy may be estimated with a fair degree of approximation, but that the surface excess free energies of high surface energy solids cannot be estimated with any reasonable degree of approximation whatever. This procedure also assumes that, within the stated limitations, θ = zero exactly; hence, that the spreading coefficient equals zero exactly.

Surface Deformation

The discussion above tacitly assumes that the surface of the solid is not deformed by the sessile drop. In Fig. 1.8c the surface tension of the liquid at the three-phase point of contact is shown resolved into both the horizontal component $\gamma_l \cos \theta$, and the vertical component $\gamma_l \sin \theta$. The latter is balanced by an equal downwards pressure at the centre due to the action of the surface tension of the envelope of the drop producing an inward pressure. The vertical component is of no significance for the contact angle equilibrium discussed above, provided that the surface is not deformed by its action. Bailey has shown that, in the case of a drop of mercury (γ_l = 515 dynes/cm) placed on a very thin slice of mica, a peripheral ridge is raised in the mica by the action of the vertical component. The high surface tension of the mercury and the extreme thinness of the mica both contributed to making the effect observable. Michaels and Dean found that oil-drops on thick layers of 4% silica gels gave constant contact angles in accordance with the assumption of an undeformed surface. It has been shown that surface deformation is negligible provided that $\gamma_l/Et \ll 1$, where E is Young's modulus of elasticity and t is the width of the annular ridge. For γ_l about 100 dynes/cm and t = 10 nanometres, E must be greater than 10^{10} dynes/cm² to validate the assumption of a surface undeformed by the vertical component of the surface tension of the drop. E varies from 16 to 390×10^{10} dynes/cm² for metals, is about 1×10^{10} dynes/cm² for common timbers and glassy polystyrene at room temperature, and is about 0.5×10^{10} dynes/cm^2 for india-rubber. It may be concluded that, except for exceptionally soft substances or extremely thin foils not anchored to a rigid support, errors due to surface deformation will be negligible.

Lester, in a paper presented at a Symposium on Wetting (Bristol, 1967), considers that not only are the elastic moduli of organic substrates of the order of 10^{10} dynes/cm^2 but also that this is so far reduced in the interfacial layers in contact with a sessile drop that the surface may be regarded as freely mobile, just as if it were liquid. He concludes that surface deformation of such solids invalidates the conventional contact angle relationships. The present writer is unconvinced that this is valid except for special cases.

Adhesives—Theoretical

In order to get good adhesion between substrate and adhesive the first essential is to ensure good *molecular* contact. Solution-type adhesives usually do ensure this since the "open time" during which the solvent is allowed to evaporate is ample to allow the dissolved or dispersed adhesive to adsorb on to the solid surface so that the individual extended or loosely coiled adhesive molecules may make many points of attachment to the surface. If the solvent does not, itself, spontaneously wet the surfaces to be joined, then provided the surfaces are permeable to the solvent or solvent vapour, forced wetting must be resorted to, the solution of adhesive being squeezed between the surfaces in such a manner as to allow the displaced air to escape. The surfaces are then held together until the solvent has evaporated and the adhesive molecules form a continuous solid structure. The same principle applies to "hot-melt" adhesives. This difficulty arises only when the solvent of hot-melt adhesive is of higher surface energy than the adherend. Once molecular contact has been ensured between the solid adherend–solidified adhesive interfaces, the question of "wetting" or "non-wetting" becomes irrelevant. Much confusion has arisen because, in many practical applications of adhesives, the time necessary to establish the requisite degree of molecular contact across the interface is not allowed. A simple experiment may be quoted by way of illustration. A piece of polythene sheet was clamped between two standard test pieces of beech-wood to form a 1 inch lap-joint and the assembly was placed in an oven at 110°C and left overnight. After removal from the oven the assembly was allowed to cool to room temperature before the clamps were removed. The joint was then tested in the Hounsfield tensometer, rupture occurring at a load of 500 kg. Although the adhesive was chemically a high melting paraffin wax, the joint did not fail in adhesion but in cohesion in the wood. On the other hand, an attempt to use polythene to make plywood from 1.6 mm veneers using a press temperature of 115°C and press times of 2, 5, and 10 min failed. The veneers separated easily under a knife test, adhesion between wood and polythene being weak under these conditions.

These sharply contrasting results with the same adhesive and substantially the same adherend, emphasizes the point that the factor of overriding importance in the practical use of adhesives is not the chemical nature of the adhesive but the formation of a "proper joint". A "proper joint" is one in which the films of air, water, or oil on the adherend surfaces have been removed and the conditions of application of the adhesive (time, temperature, pressure) have been such as to permit the formation of true molecular contacts between the adherend and the adhesive over the greater part of the surface area to be joined.

Foams

Perhaps the most widely appreciated property of surface active substances in aqueous solution is their ability to promote the formation of foams and bubbles. The presence of foam on the surface of a washing solution has long been regarded as an indication that the solution contains an adequate amount of washing agent; every child is familiar with the rapidly changing colours of a soap bubble. Industrially, protein foams are used extensively in fire fighting in the open and are particularly valued for blanketing petrol fires. Detergent solutions have been developed to form complete "plugs" of foam which may be carried by the ventilation air current along coal-mine roadways to the seat of a fire and put it out. The method was devised and demonstrated by the Safety in Mines Research Establishment at Buxton in Derbyshire, and has been successfully applied to deal with a major fire outbreak in an actual mine in the United States, but has not been adopted by the British coal industry. The separation after grinding of the rock of particles of metallic ore from particles of country rock by the "flotation" process is a further application of foam formation of immense industrial value. The constructional value of foamed concrete has been thoroughly established, and its use is rapidly growing.

A freely floating bubble is spherical since this shape has the lowest surface–volume ratio: the total solution–air surface, however, is very large in comparison with the surface of a single drop of liquid containing the same volume of water as the thin film forming the bubble. Hence although the surface energy of a thin film is a minimum when the film is in the form of a spherical bubble, the surface energy of the bubble is high in comparison with a drop containing the same volume of water. Thus the bubble is, at best, only in a state of meta-stable equilibrium. This state can be maintained for substantial times mainly because of the existence of an elasticity E which is proportional to the area A of the film and the rate of change of surface tension with change in area ($d\gamma/dA$). That is, the more the surface tension increases with a sudden extension of the film, the greater the restoring force opposing the extension of the film. According to Gibbs,

$$E = 2A\, d\gamma/dA = 2\, d\gamma/d \ln A. \qquad (1.38)$$

Vertical films drain, the rate of drainage following Poiseuille's law for viscous flow. The two surfaces thus approach each other and the film, at a given horizontal plane, runs through the series of thin-film interference colours until it becomes too thin to exhibit any interference at all—the "black" film. Black films have an ordered lamellar structure, changing in thickness by sudden steps each of which is of a height corresponding to the length of a single molecule of the tenside. The thinnest possible film consists of a bimolecular leaflet of tenside with only a few molecular layers of water between them.

It has been suggested that the thinnest black films may spontaneously rupture due to random thermal motion of the film molecules producing a hole in the film of a molecular diameter or so. Certainly if a hole were to be formed by such a mechanism, the tension in the surrounding film would rapidly enlarge it. It is, however, interesting to note that spark photographs exhibited by Quaille in the late 1920s of the passage of a rifle bullet through soap bubbles showed that it was the exit hole which enlarged to burst the bubble, whereas the inlet hole tended to heal. The bubbles were not black films of minimum thickness. The preservation by Dewar of black films of large area for as long as 2 or 3 years shows that the probability of rupture by such a statistical fluctuation process is vanishingly small. In many practical foams it may readily be observed that rupture of bubbles begins while the water films within them are still thick—even too thick to show interference colours. Rupture by hole formation by statistical fluctuation in such instances is an even more remote possibility.

Carboxylate soap films are more stable than those of pure non-soap anionic tensides unless sufficient alkali has been added to the solution to suppress the hydrolysis. Fatty acid produced by hydrolysis is almost completely insoluble in water. In tends to be dissolved into the soap micelles and to be strongly adsorbed on the surface so that, although only a small fraction of the soap is hydrolysed (about 0.2%), the surface layer contains as many un-ionized fatty acid molecules as fatty acid anions. The adsorbed layer of molecules, instead of forming a "gaseous" adsorbed film as with pure non-hydrolysing tensides, forms a "condensed" film of high viscosity or even plasticity. This property may be introduced into the sulphate and sulphonate classes of tensides by the deliberate addition of suitable insoluble, or slightly soluble, substances such as fatty alcohols, amides, ethanolamides, sulphonamides, or ureides.

Thin films formed with ionizing tensides, on thinning, bring two similarly charged layers towards each other. Mutual repulsion between the charges tends to hinder the process of thinning. Thin films formed by non-ionizing tensides lack this resistance to continued thinning. Further, their films appear to have a much lower elasticity than those of soaps and anionics. In general, therefore, solutions of non-ionics tend to form weak, unstable, foams. There is, however, a narrow range of ratio of polyether chain to hydrocarbon within which the foaming power of non-ionics is fairly high, although less than that of the anionics.

The gas within a spherical bubble is at a higher pressure than the external pressure. Owing to the curvature, the surface tension in the film has an inwardly directed component, hence the greater the curvature the greater the excess internal pressure. The value of the excess pressure is easily calculated.

The work done in the expansion of the bubble from radius r to $r+dr$ is balanced by the increased free surface energy due to the increase in area dA, i.e.

$$dP \, dV = \gamma \, dA$$

but $$dV = 4\pi r^2 \, dr \quad \text{and} \quad dA = 16\pi r \, dr.$$

(Since there are two liquid–gas surfaces, the formula for the area of the bubble must be doubled.) Thus

$$dP \, 4\pi r^2 \, dr = \gamma \, 16\pi r \, dr$$

or $$dP = 4\gamma/r. \tag{1.39}$$

(For a bubble within a liquid, $dP = 2\gamma/r$.)

A foam is a collection of bubbles in contact, ideally forming a structure of contiguous dodecahedra to which real foams approximate. The foam films are flat between neighbouring gas cells of the same size, but at the lines of three-plane contact have a spherically triangular cross-sectional region (Fig. 1.10). Since the radius of curvature is negative, this region is of lowered pressure, and hence liquid tends to be sucked into these borders. In a non-uniform foam the intervening liquid walls are no longer flat but curved, and the gas pressure is greater in the smaller bubbles. The gas will therefore tend to diffuse through the walls from the smaller to the larger bubbles, a clearly autocatalytic process. Small bubbles will disappear with time, and the foam will tend towards a structure of nearly uniform large bubbles.

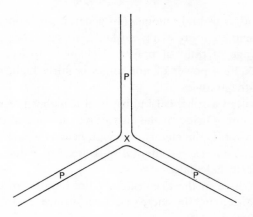

FIG. 1.10. The plateau border. Because of negative curvature at surfaces with respect to X, the pressure at X is less than the pressure at P; liquid is therefore sucked into the X region.

Substances which spread on water to films with no surface elasticity are strong foam breakers. Silicone greases and perfluoro alcohols are good examples.

Emulsions and Creams

When an immiscible pure second phase is injected into pure water in a fine jet, it breaks up into small volumes which more or less rapidly rise to the surface, surrounded by an aqueous film which rapidly drains and ruptures, permitting the temporarily discrete small volumes of second phase to re-form into a homogeneous bulk phase. If, however, a tenside is present in the system, the rupture of the aqueous layer between two bubbles of air or two drops of oil may be long delayed.

If the injected phase be an oil of density near that of water, the rate of rise of the drops to the surface ("creaming") may be very slow and the numerous oil-drops are relatively stably dispersed as an emulsion. If the injected phase be of low density compared with water (e.g. air) the rate of creaming of the drops will be high and the "emulsion" state will be evanescent. The "creamed" state, i.e. the foam, may however, persist for a long time. From the point of view of systematics there is no essential difference between a foam and a creamed emulsion: although the numerical values of important parameters may differ widely in the two cases, no essential differences of principle are involved.

The interfacial free energies solution–air and solution–oil may be lowered to 25–30 ergs/cm^2 and 6–10 ergs/cm^2 respectively. In special cases the solution–oil interfacial free energies may be lowered still further, and at values of 0.1 ergs/cm^2 or less, spontaneous emulsification will set in. This type of behaviour does not occur at the solution–air interface.

Davies and Haydon show that the mechanism first proposed by Gurwitsch and later by Raschevsky is able to explain many cases of spontaneous emulsification, e.g. ethanol in toluene or water. *Diffusion* of the mixture into water occurs with consequent steady dilution of the alcohol until the toluene is "stranded" as fine emulsion droplets. The process may be accompanied by violent surface turbulence which has itself been held responsible for spontaneous emulsification. But a similar system, in which the turbulence is completely inhibited by a little protein in the water, continues to emulsify spontaneously by the "diffusion and stranding" mechanism. Emulsification by surface turbulence may occur if the system is one of exceptionally low interfacial tension (< 0.05 dyne/cm). However, acetone in toluene on water shows vigorous surface turbulence but no emulsification.

Virtually negative interfacial tensions lead directly to spontaneous emulsification. Interfacial tension versus log c curves for cetyl alcohol in toluene on buffered sodium decyl sulphate solutions of increasing concentration are rectilinear down to the zero γ-axis. At sodium decyl sulphate concentrations from slightly less than the concentration at which the graph meets the zero interfacial tension axis, to all higher concentrations, spontaneous emulsification is observable.

In the general case, the emulsion droplets will vary continuously over a wide size range and the determination of the particle size distribution is often of great technical importance. At a certain time after the prepared emulsion has come to rest, drops of successive size ranges, originally evenly distributed from top to bottom, will have completely cleared below horizontal planes at increasing heights from the base as the radii of the drops increase in size. Several means are available for the estimation of the particle size distribution which depend on this simple principle.

The most obvious is to remove successive layers of the emulsion at a given moment and analyse each layer for its content of the dispersed phase. As often happens, the "simplest" method is difficult in practice because of the ease with which the system is disturbed and layers partially mixed by convection when it is attempted to draw off the successive layers.

Sedimentation of heavier-than-water particles may be measured by setting the system up with a balance-pan immersed in a plane near the base. The curve given by plotting the accumulated weight of sediment in the pan against the time may be graphically differentiated so as to yield the particle weight distribution. In principle, the same method could be applied to an emulsion creaming upwards, the pan lying in a plane near the upper surface and the drop accumulation being held to the underside of the pan by provision of a slight concavity.

The change in distribution with height over a period of time leads to a fall in the centre of gravity. This may be detected by a hydrometer floating in the emulsion or by a sensitive manometer attached to a side-tube. An ingenious application of this principle was made by Manning and Taylor in 1936 to solve the problem of measuring the rate of settling of colloidal coal dispersed in fuel oil in the bunkers of oil-fired ships. The dispersion was placed in a test-tube fitted with a collar carrying knife edges which rested on polished planes. Initially, the tube was filled so that the centre of gravity lay only about a millimetre below the plane of suspension. The period of the system as a pendulum was measured initially and at intervals of several hours. The period is extremely sensitive to small changes in the centre of gravity. The behaviour of the fuel in the bunkers in the course of many months could be predicted in a few days.

As commonly used, the term "stability" is ambiguous in its application to emulsions. In one sense it connotes stability to creaming, in another stability to coagulation, and in yet a third sense stability to coalescence of the drops. Creaming is a tendency of the drops to concentrate in the upper or lower layers of the emulsion. It depends on the drop sizes, the difference between the densities of the two phases, and the viscosity of the external phase. Coagulation is the tendency of drops to stick together on contact whilst retaining their individuality. It depends on the number of drops per unit volume (which governs the collision frequency) and on the kind of stabilizing film at the interface between the drop and the external phase (which governs the proportion of collisions which result in mutual attachment of the colliding drops). Coalescence is the fusion of two or more drops into one larger drop due to breakdown of the interfacial films separating the drops, and leads finally to the formation of a free bulk oil layer. Coagulation must always precede coalescence.

The film of aqueous solution between the oil drops in a coagulum or cream is subject to drainage in the same way as are the films of water between the gas bubbles in a foam. Coalescence occurs when the aqueous film breaks down. As with foams the breakdown may occur before the film has drained to extreme thinness.

The rate of creaming of dilute emulsions is, of course, governed by Stokes's law which relates the rate to the particle size, the density difference between the two phases, and the viscosity of the dispersion medium,

$$v = 2\pi r^2 g(\varrho_1 - \varrho_2)/9\eta, \tag{1.40}$$

where v is the velocity of the emulsion drop of radius r, $(\varrho_1 - \varrho_2)$ is the density difference between the phases, and η is the viscosity of the external phase.

The emulsion drops are in random motion. The number of collisions per second will be proportional to the square of the number of drops per cubic centimetre. The proportion of collisions which end in mutual attachment will be a constant for the given system. By convention, an aggregate of drops remaining attached is counted as one drop. The rate of coagulation is then a function of the rate at which the number of drops per cubic centimetre decreases with increase in time.

The rate of change in the number of drops is given by

$$\left.\begin{array}{c} dn/dt = -kn^2 \\ = -Afn^2, \end{array}\right\} \qquad (1.41)$$

where A is the collision frequency constant and f is the fraction of the collisions which are inelastic.

Hence $\qquad\qquad n = n_0/(Aftn_0 + 1), \qquad\qquad (1.42)$

where n is the number of drops after time t and n_0 is the initial number of drops per cubic centimetre.

Coalescence, i.e. breakdown of the films separating the drops, is a random process (as is also true for foams). The rate of coalescence for a creamed emulsion in which the drops are separated only by moderately thin to thin aqueous layers is therefore a function only of the number of films between drops; hence it follows a first-order law. The rate of appearance of free oil, however, will depend on the relative values of the rate of coagulation and the rate of coalescence. If the former is large compared with the latter, it will follow first-order kinetics, and if coalescence is rapid in comparison with coagulation, it will follow second-order kinetics.

Well-stabilized emulsions may change in particle size distribution because of the definite, even if very small, solubility of the inner phase in the outer phase. The "solution pressure" of the smallest drops will be greater than that of larger drops. It was shown in the preceding section on foams that the pressure within a bubble or drop with a single interface is greater than that of the external liquid in the same plane by an amount given by the expression $\Delta P = 2\gamma/r$, where γ in this context is the oil/solution interfacial tension. If γ is put $= 10$ dynes/cm and two drops of radii 10×10^{-4} cm ($10\ \mu$) and 0.1×10^{-4} cm ($0.1\ \mu$) respectively are considered, then the two pressure differences are 20×10^3 dynes/cm and 20×10^5 dynes/cm respectively.

Increase of external pressure on a liquid connotes increase in the vapour pressure; hence, by Henry's law (since the aqueous solubilities of the internal phase will be small), it also connotes a proportional increase in solubility with decreasing droplet size. The smallest drops therefore tend to disappear by transfer through the external phase to the larger drops. Harkins and co-workers carried out painstaking size-frequency studies by statistical treatment of extensive microscopic measurements in the early 1930s, and showed that droplet size increased over a period of several months until the adsorbed layer at the surface became condensed. Lawrence and Mills carried out similarly detailed and precise investigations in 1954 and showed that the average volume of the drops increased linearly with time.

Also in the early 1930s, Harkins and Robinson independently published a simple theory of the electrical stabilization of emulsion droplets. Consider the approach under random motion of two droplets. If the adsorbed film of ionized tenside molecules is "gaseous", then, as the droplets approach, electrostatic repulsion will tend to drive the adsorbed molecules away from the opposing surfaces of the droplets to the rear surfaces and the particles may touch at the bared surfaces and coalesce. If the adsorbed film is close-packed, this process cannot occur, and the opposed surfaces of the approaching droplets are not only coated with strongly water-attracting groups which will resist the expulsion of water from between the drops but are also highly charged, leading to an additional strong electrostatic repulsive force between the two drops.

Ionized tensides alone, of the chain lengths normally used, rarely, if ever, form condensed films at aqueous–oily interfaces at the temperatures at which technical emulsions are normally made, stored, and used. The adsorption film may be condensed by the addition of a second non-ionizing type of tenside. It might be thought that, even with a liquid condensed film obtained in this way, it would still be possible for the ionic tenside to move round to the backs of the approaching droplets, and, to some extent, this must be true. However, such separation of the ionic and non-ionic tensides will be opposed by the work required to separate the molecules against the ion-dipole attractive forces between molecules of the two kinds and by the decrease in entropy consequent on such an "unmixing" process. In practice, mixtures of ionic and non-ionic emulsifiers afford highly satisfactory emulsifying agents.

Non-ionic tensides alone readily form close-packed surface films. Their hydrophilic groups will strongly oppose the expulsion of water from between approaching drops.

It must not be forgotten that water-soluble polymeric substances such as the gums, mucilages, saponins, and proteins may also be excellent emulsifying agents. Proteins, in particular, are the emulsifying agents in naturally occurring emulsions such as mammalian milk and the lattices of rubber-yielding plants. In many such instances, and with emulsions stabilized by mixtures of water-soluble and oil-soluble emulsifiers, the interfacial film may become plastic and tough. Further, powdered solids which are partially wetted by both phases so that at a plane surface of the solid the two liquids will reside with a finite angle of contact, will stabilize an emulsion. The liquid which wets the solid most usually forms the external phase.

When two pure, virtually immiscible, liquids are violently agitated together, both phases are broken up. At the moment agitation ceases the system is a mixture of oil in water and of water in oil: the system rapidly separates again into two phases. In the presence of an emulsifying agent the rates of breaking of the two types of emulsion will usually be widely different and will depend on the nature of the emulsifier; the system rapidly settles into the emulsion type with the lower rate of breakdown.

If the rate of breakdown of the oil-in-water emulsion is given by $rate_1$, and that of the water-in-oil emulsion by $rate_2$, then, as shown by Davies,

$$\text{rate}_1 = A_1 \exp\left[-\Sigma \Delta E_1 / RT\right],$$
$$\text{rate}_2 = A_2 \exp\left[-\Sigma \Delta E_2 / RT\right], \tag{1.43}$$

where A_1 and A_2 are the collision frequencies and $\Sigma \Delta E_1$ and $\Sigma \Delta E_2$ are the sums of the various energy terms in the energy barrier to coalescence of two contacting drops due to electrical repulsions, hydration energies of adsorbed molecules, and so on. If the phase volumes are equal the collision frequencies, A_1 and A_2 may be assumed equal and

$$\text{rate}_1/\text{rate}_2 = \exp\left[(\Sigma\Delta E_2 - \Sigma\Delta E_1)/RT\right]. \qquad (1.44)$$

Davies estimated the energy terms for characteristic chemical groups, e.g. $-CH_3$, $-CH_2-$, $-OH$, $-COOH$, $-O-$, etc., and was able in this way to put into quantitative form "Bancroft's rule" that the phase in which the emulsifier is the more soluble will form the continuous phase, and Griffin's empirical HLB scale in which low values are assigned to water-in-oil emulsifiers and high values to oil-in-water emulsifiers. HLB stands for hydrophil-lipophil balance. Bancroft's rule, however, describes a common tendency only and is subject to a number of exceptions. In particular, lower ethylene oxide condensates of higher fatty alcohols, although far more soluble in oils than in water, are excellent oil-in-water emulsifiers. A product of this class has been proposed by Imperial Chemical Industries, Ltd. for the dispersal of crude-oil slicks at sea. Much smaller amounts are required than of water-soluble dispersants for this purpose; the treated oil does not adhere to the feathers of sea birds or other surfaces; the dispersant does not pass into the surrounding sea water and is of low toxicity to marine life.

Non-ionic tensides are highly soluble in oils, even when the proportion of ethylene oxide combined with the paraffinic part of the tenside is high. Those with few $-O \cdot CH_2CH_2-$ groups are of slight or negligible solubility in water; their partition coefficients C_{oil}/C_{water} are extremely large. Those with many $-O \cdot CH_2CH_2-$ groups are readily soluble in water as well as in oils, and their partition coefficients are moderate—of the order unity. The former have been assigned low HLB numbers and the latter moderate to high HLB numbers. Table 1.5 lists typical tensides, their HLB numbers, and their characteristic area of usefulness, as given by Griffin.

It is often more efficient to use a mixture of tensides of widely differing HLB numbers, giving the desired values by proportional parts, than to use a single tenside which has just the required HLB value.

However, this is not the whole story of the factors governing the type of emulsion formed, water-in-oil or oil-in-water. In the laboratory emulsions will usually be prepared in glassware. Industrially, they may be prepared by feeding the aqueous and oily streams into a narrow

TABLE 1.5. TENSIDES AND THEIR HLB VALUES

Tenside	HLB value	Characteristic use
Sorbitan tetrastearate	0.5	
Oleic acid	1	
Cetyl alcohol	1	
Span 85 (sorbitan trioleate)	1.8	
Sorbitan tristearate	2.1	
Arlecel C (sorbitan sesquioleate)	3.7	Water-in-oil emulsifiers: optimum = 5
Octyl phenol-1 ethanoxide	4.0	
Span 80 (sorbitan monooleate)	4.3	
Sorbitan monostearate	5.9	
Sorbitan monopalmitate	6.7	Wetting agents
Span 20 (sorbitan monolaurate)	8.6	
Octyl phenol-4 ethanoxide	9.6	Oil-in-water emulsifiers: optimum = 10
Tween 81 (sorbitan monolaurate-6 ethanoxide)	10.0	
Triethanolamine oleate	12	Detergents
Octyl phenol-10 ethanoxide	14	
Tween 80 (sorbitan monolaurate-20 ethanoxide)	15	Solubilizers
Potassium oleate	20	

space between two surfaces of which one may be fixed while the other is rapidly rotated. The two fluids are violently mixed together with intense shearing action, and may undergo a single pass or be re-cycled. In studying emulsification in a laboratory-scale version of the industrial process, Davies found that the emulsion type depended not only on the kind of emulsifiers used but also on the material of which the shear plates were made. If the plates were more readily wetted by the aqueous stream, oil-in-water emulsions were readily formed: if they were more readily wetted by the oily stream, then water-in-oil emulsions were more readily formed. A still further factor was the relative volumes of the two phases fed into the shear-plate gap.

HLB values can be assigned to such small molecules as methanol (HLB = 8.3), ethanol (HLB = 7.9), butyl acetate (HLB = 10), and so on. Extended in this way the system ceases to have any useful function. Any emulsion formed by these small molecules would be too unstable to be of practical value. The HLB system, therefore, is of use only as a guide to the type of emulsion which will be the more favoured, but says nothing about the likely stability to coalescence of the emulsion. Experience shows that it is necessary to use an agent with a hydrocarbon chain of at least twelve carbon atoms, or the equivalent ring systems (benzene = $4CH_2$), and a water-solubilizing group sufficient to raise the HLB value to that required in the particular application under consideration. The author of the HLB system was explicit in limiting the application of the system to substances of the requisite molecular weight range; some other workers have been less cautious.

Detergency

The general problem of detergency is very broad. It ranges from washing the hands with a toilet soap to removing mustard gas (a heavy grease in spite of its name) from roads and buildings, or radioactive fall-out from all kinds of exterior surfaces. It covers alike the washing of fine fabrics and the scouring of raw woollen fleeces, the removal of milk-stain from glass bottles, egg residues from crockery, and bloodstains from surgeon's overalls. Clearly no one all-embracing theory of detergency is possible: the removal of oils and heavy greases, of adherent soiling such as proteins, and of finely divided carbonaceous or clayey particles from a variety of surfaces, each requires individual treatment.

Proteins may be solubilized by a change of pH or by the action of neutral salt solutions, milk-stains on glassware by strongly alkaline solutions, for example of caustic soda or sodium metasilicate, and egg residues on crockery largely by mechanical scraping with a dish-

5*

cloth. Domestic detergent powders have been marketed containing mixed enzymes to digest proteins, starches, and fats from fabrics.

Exterior surfaces may be cleaned of radioactive heavy-metal contamination by hosing down with solutions of sequestering agents, transferring the problem of disposal to the sewers, sewage purification plants, and streams. Heavy-metal complexes of ethylene diaminetetra-acetate, for example, are extremely water-soluble.

Tensides become the agents of choice for dealing with oils, greases, and finely divided particulate matter. Adsorption of tensides at the oil–water interface and at the solid–solution interface lowers the interfacial tensions and hence markedly influences the contact angles. Generally, the adhesion of the aqueous solution to the interface is increased relative to that of the oil or grease, and the water therefore tends to spread preferentially, the oil or grease rolling up into spheres if small in amount, but in the case of heavy soiling, gravitational (buoyancy) and frictional (viscous drag) forces may draw out thick threads of grease which break up into a number of drops by the action of surface tension.

Two mechanisms have to be considered in the case of solid dirt. If the solid be a polar fatty substance at, roughly, not more than 20°C below its melting point, then, as Lawrence has shown, the edges and surfaces of the crystalline solid matter take up both detergent (which concentrates at the surface by adsorption) and water to form a ternary, liquid crystalline phase, which usually grows out from the crystal surfaces in microscopic twisting tubes of concentric molecular lamellae, which resemble the sheaths of nerve fibres and are hence called "myelinic forms". As viscous phases they may be dispersed into the bulk of the aqueous solution.

Particulate dirt most commonly consists of microscopic particles of clay, carbon, and silica. Often they occur on the surface more or less embedded in or surrounded by oil or grease. If the solid surface is woollen fabric it is sufficient to remove the grease: the particulate dirt is removed in direct proportion to the grease removed. If the solid surface is cotton fabric, this simple relationship no longer holds: it is possible to remove all the grease and leave behind all the particulate dirt. More generally, some particles come away with the grease, but by no means all. Worse, it is only too easy to remove the particulate dirt only to redeposit some of it, leaving the fabric apparently dirtier than before the cleaning started.

The resultant force between surface and particle in plain water is a complex function of distance of separation. Close to the surface there is a deep potential energy well. Somewhat further out there is a potential energy "hump", and yet further out still there is a relatively shallow potential energy minimum, the "secondary" minimum. The theory is complex and has been developed over many years by Verwey, Hamaker, Overbeek, Dervichian, Derjaguin, Casimir, Polder, and many others. Particles in the secondary minimum tend to linger in it, and hence particles tend to collect together in this region. The deeper the secondary minimum the stronger the tendency for the particles to collect or flocculate together: few particles will leave the floccule to pass into the bulk water, few will climb the "hump" and drop into the deep well very close to the surface from which escape at ordinary temperature (up to and including the boil) is impossible. The calcium and magnesium salts in "hard" water have the effect of deepening the secondary minimum.

Tensides of the detergent class, particularly in the presence of "builders" which precipitate or sequester the calcium and magnesium ions, tend to reduce or eliminate the secondary minimum, and may raise the height of the hump. Thus the immediate effect of the detergent solution is to disperse the dirt floccules to individual particles scattered throughout the solution; the fabric is cleaned.

The mere fact of this dispersion means that many more particles than before are on the slopes of the hump; some of these will acquire sufficient kinetic energy in the right direction to pass over the top and drop into the deep well. The inner slopes are now too high and too steep for any particle to climb; passage into the well is therefore a one-way traffic. The process is slow but the final state is that the dirt, originally loosely attached in occasional and readily dispersible clumps, is evenly and irremovably distributed as individual particles over the entire surface of the fabric which now looks dirtier than before. This effect is strongly inhibited by minute traces of carboxymethylated cellulose which has the effect of raising the hump in the potential energy curve still further and hence of slowing down markedly the rate at which particles drop into the well.

Water Conservation

In hot countries the loss of water from lakes and reservoirs is large and of serious social and economic importance. Insoluble fatty compounds carrying terminal water-attracting groups such as —OH, —COOH, —CONH$_2$, spread from the peripheries of crystals lying on a water surface to form a unimolecular surface film. Depending on the temperature and the precise nature of the compounds, the surface film may have the property of a two-dimensional gas, vapour, liquid, or solid. The condensed, liquid or solid, films cover the water surface so completely that they substantially retard evaporation of the water.

Cetyl alcohol, C$_{15}$H$_{31}$CH$_2$OH, has been successfully exploited commercially, particularly in Australia, where the retention of upwards of a 6 ft layer of water per annum over large lakes by retardation of solar evaporation owing to the presence of a monolayer of cetyl alcohol has increased the volume of water available for use by many millions of tons at a trivial cost.

Reading List

The reader who wishes to pursue the subject more deeply will find the following useful.

General Works

ADAM, N. K., *Physics and Chemistry of Surfaces*, 3rd edn., OUP, 1941.
DAVIES, J. T. and RIDEAL, E. K., *Interfacial Phenomena*, 2nd edn., Academic Press, 1963.
DANIELLI, J. F., PANKHURST, K. G. A., and RIDDIFORD, A. C., *Surface Phenomena in Chemistry and Biology*, Pergamon Press, 1958.

Intermolecular Forces

Intermolecular forces, *Discussions of the Faraday Society*, No. 40, 1966.

Solubility

HILDEBRAND, J. H. and SCOTT, R. L., *The Solubility of Non-Electrolytes* (reprint of 3rd edn.), Dover Publications, 1950.
The equilibrium properties of solutions of non-electrolytes. *Discussions of the Faraday Society*, No. 15, 1953.

Tensides

MOILLIET, J. L., COLLIE, B., and BLACK, W., *Surface Activity*, 2nd edn., Spon, 1961.
DURHAM, K. (ed.), *Surface Activity and Detergency*, Macmillan, 1961.
Proceedings of the International Congresses on Surface Activity:
    Paris, 1954, Chambre Syndicate Tramagras.
    London, 1957, Butterworth.
    Cologne, 1960, Deutsche Ausschuss für Grenzflächenaktive Stoffe.
    Brussells, 1964, Gordon & Breach, London, 1965.
    Barcelona, 1968 (to be published).

Foams

MYSELS, K. J., SHINODA, K., and FRANKEL, S., *Soap Films*, Pergamon Press, 1959.
DE VRIES, A. J. *Foam Stability*, Rubber Stichting, Delft, 1957, No. 326.

Adhesion

Wetting, Society of Chemical Industry Symposium, 1967.
ALNER, D. J. (ed.), *Aspects of Adhesion*, ULP, Vols. 1, 2, 3, and 4.

CHAPTER 2

Soap and Detergents

Introduction

The origin of soap-making is unknown. The Phoenicians were acquainted with it by at least 600 BC and it was known to the Gauls not later than about 300 BC. It was the Gauls who taught the art to the Romans whose well-preserved soap factory was recovered at Pompeii complete with still usable soap tablets. The early method of production, by heating goat's tallow with wood-ash (crude potassium carbonate) and conversion of this resulting soft soap to hard soap by salting-out with sodium chloride brine, suggests an origin among goat-herders in well-wooded country rather than among the more sophisticated cultures of the relatively treeless regions of Egypt and Babylonia. Its original use seems to have been as a shampoo and cosmetic. In the Western world its use for personal cleansing made only slow progress between 600 BC and AD 1600. Its medicinal properties were already known to Galen.

Soap-making

Ordinary soaps are blends of alkali–metal or organic base salts of fatty acids of from eight to eighteen carbon atoms, the essential feature being that the salts must be soluble in water at temperatures less than 100°C, usually much less. Hard soaps are sodium salts, soft soaps are potassium salts, shaving soaps usually blends of sodium and potassium salts of the aliphatic acids C_8–C_{18}. Ammonium, mono-, di-, and triethanolammonium soaps are used in the cosmetic field. A portion of the fatty acids in the soap blend may be replaced by rosin—mainly abietic acid, a tricyclic terpene $C_{20}H_{30}O_2$—

(numbering as a tricylic terpene)   (numbering as a steroid)

Abietic Acid

The fatty acids are mainly derived from natural glycerides, rarely from waxes other than spermaceti (palmityl palmitate). The major sources are from the seeds and fleshy fruits of plants and the body fats of land and marine animals. Some typical fatty acid compositions of a number of common fats are given in Table 2.1. The figures quoted are not to be taken too literally: quite wide variations occur from season to season for any one species, between varieties within species, between crops taken from different soils, and so on. The major components may vary by as much as $\pm 15\%$, minor components from about half to about double the quoted figure. Notwithstanding this variability, broad separation into easily recognizable classes can be made. For details, the classic work of Hilditch should be consulted.

The pure fats are triglycerides, the fatty acids being distributed at random over the available sites. The highest quality fats are largely used for edible purposes. The lower quality fats used for soap-making have often undergone some enzymic degradation leading to the presence of free fatty acids, diglycerides, and monoglycerides. These components are not entirely disadvantageous as the monoglycerides and the soap, formed initially by reaction of the free fatty acids with the first amounts of caustic alkali used, materially assist in the emulsifica-

TABLE 2.1. TYPICAL FATTY ACID COMPOSITIONS OF SOME COMMON FATS

Source of fat component	→	Coco-nut	Palm kernel	Palm fruit	Olive	Ground nut	Cotton seed	Sun-flower seed	Mut-ton tallow	Beef tallow	Lard	Whale	Her-ring
Fatty Acid:													
Saturated													
Caprylic	C_8	8	3	—	—	—	—	—	—	—	—	—	—
Caproic	C_{10}	7	5	—	—	—	—	—	—	—	—	—	—
Lauric	C_{12}	46	47	—	—	—	—	—	—	—	—	0.2	—
Myristic	C_{14}	15	15	2	0.4	—	—	—	2	5	3	7	7
Palmitic	C_{16}	9	7	42	18	6	20	6	25	28	25	19	13
Stearic	C_{18}	2	2	4	—	3	2	2	27	21	13	2.4	0.5
Arachidic	C_{20}	—	—	—	—	6	—	—	—	—	—	—	—
Behenic	C_{22}	—	—	—	—	—	—	—	—	—	—	—	—
Mono-unsaturated													
Tetradecanoic	C_{14}	—	—	—	—	—	—	—	—	—	—	—	0.7
Palmitoleic	C_{16}	—	—	—	—	—	—	—	—	—	—	2	6
Oleic	C_{18}	6	14	42	68	60	24	25	41	40	45	13	—
Eicosenoic	C_{20}	—	—	—	—	—	—	—	—	—	2	38	—
Di-unsaturated													
Linoleic	C_{18}	1	1	9	12	25	50	66	4	3	15	—	20
Eicosadienoic	C_{20}	—	—	—	—	—	—	—	—	—	—	—	28
Docosadienoic	C_{22}	—	—	—	—	—	—	—	—	—	—	—	23
Tri-unsaturated													
Eicosatrienoic	C_{20}	—	—	—	—	—	—	—	—	—	—	13	—
Tetra-unsaturated													
Docosatetraenoic	C_{22}	—	—	—	—	—	—	—	—	—	—	6	—

tion of the remainder of the fat. Since saponification occurs only at the boundary of the initially immiscible aqueous and fatty phases, extension of the interfacial boundary area by emulsification greatly promotes the reaction. Chemically, saponification is a straightforward process of alkaline hydrolysis of an ester:

$$
\begin{array}{l}
CH_2OOCR_1 \\
| \\
CH\!-\!OOCR_2 + 3H_2O \xrightarrow[\text{OH' catalyst}]{} \\
| \\
CH_2\!-\!OOCR_3
\end{array}
\qquad
\begin{array}{l}
CH_2\cdot OH \quad R_1COOH \\
| \\
CH(OH) + R_2COOH \\
| \\
CH_2\cdot OH \quad R_3COOH
\end{array}
$$

$$3\ RCOOH + 3\ NaOH \longrightarrow 3\ RCOONa + 3\ H_2O$$

However, the physics of emulsification dominates the practical reaction rate.

Some oils contain di- and poly-unsaturated acids such as linoleic acid (octadeca-8, 11 dienoic acid), linolenic acid (octadeca-8, 11, 14 trienoic acid), and similar C_{20} and C_{22} acids, as is shown in Table 2.1. These are not suitable for soap-making because of their strong tendency to spoilage by oxidation which leads to the typical odour of rancid fats and of the "drying oils" used in paints. Partial hydrogenation (finely divided nickel catalyst), which can be controlled to attack the polyunsaturated acids preferentially, converts the highly unsaturated fats to less unsaturated, eventually to saturated, fats, acceptable for soap-making. The mean unsaturation of a sample of fat is determined by finding the weight of iodine in milligrams taken up by 1 g of fat by reaction of the double bonds under specified conditions—the "iodine value".

The most common fatty acids are oleic, palmitic, and stearic. Coconut and palm-kernel oils are exceptional in containing major proportions of combined lauric acid and appreciable amounts of combined myristic acid. These are valuable because of the ready solubility of their sodium soaps in cool water.

Soaps made from individual fats are no longer regarded as satisfactory. Pure tallow soaps are hard and only slightly soluble in any but very hot water. The old-time music-hall joke about the tablet of soap which, issued to the raw naval recruit, lasted until he retired as an admiral, refers to a typical tallow soap. The nineteenth-century Marseilles soap, made from olive oil, was too soft and too easily soluble, but was widely regarded as notably mild to the skin. Soaps made from coconut or palm-kernel oils are over brittle although dissolving fairly well even in sea water. In the manufacture of modern soaps, the various advantages of the different fats are largely retained and their disadvantages largely mitigated by appropriate blending of the various fats and oils.

Some vegetable fats are strongly coloured. The colour is removed (to a large extent) by treatment of the oil, free from fatty acid, with sulphuric acid and fuller's earth (SAFE process). An acid-activated fuller's earth may be used in place of untreated fuller's earth, no further sulphuric acid then being used. Liquid fat is stirred at 110°C and treated with 200 lb of fuller's earth and 0.5 lb of sulphuric acid per ton of fat. After 2 hr it is pumped through filter presses at about 100°C to remove the earth.

If the oil or fat contains appreciable amounts of free fatty acid, this process cannot be used since the fatty acid would be strongly adsorbed on the fuller's earth and hinder its decolourizing action as well as being lost for soap-making. Acid oils are bleached by treatment with chlorine dioxide, released into the oil by the action of the acid on sodium chlorite. A preliminary wash with hot 5% brine removes some of the impurities. The temperature of the oil is adjusted to 80°C and 0.5% of a 25% sodium chlorite brine is added, preferably in aliquots. Agitation with air is maintained throughout, and spot tests are taken from time to time to check the degree of bleaching. When the desired colour level has been reached, agitation is stopped, the mixture is allowed to separate, and the liquor is run off.

The traditional method of soap-making, in large tanks carrying up to 150 tons in each, was an empirical and highly skilled art. Caustic soda is a highly efficient coagulant in emulsion systems, hence an excess will strongly flocculate the emulsified fat, reducing the area of contact between the fatty and the alkaline aqueous phases and hindering the process of saponification. On the other hand, if the electrolyte content of the aqueous phase falls too low, for example by approaching exhaustion of the caustic soda, the soap formed may absorb all the aqueous phase and form an intractable gel. The soap-boiler's art lay in maintaining an optimum balance between excessive "graining out" (flocculation) and excessive "closing" (gel formation) of the soap throughout the saponification process. Towards the end of the saponification of the batch of fats, the amount of caustic soda required to complete the conversion of fat to soap will become insufficient to maintain the optimum balance between "graining" and "closing" the soap. From this point to the completion of the batch, sodium chloride must be added from time to time.

This traditional "open-kettle" method of soap-making is being rapidly displaced by continuous processes, of which several have been developed. These processes yield much larger throughput, occupy much less space, and make much smaller demands on power, than the open-kettle process. Details of these processes may be found in the literature cited at the end of the chapter.

Non-soap Detergents—Introductory

Down to about 1850 soap remained the only organic detergent manufactured. At about this time the first sulphonated oils were made by reacting castor oil with sulphuric acid. Sulphuric acid reacts with the hydroxyl group of the ricinoleyl (12 hydroxy-9 octadecenoic acid) radical to form the sulphate. The products, of various degrees of sulphation, were known generically as the Turkey red oils because of their use as wetting agents in the dyeing of carpet wool with madder (natural alizarin or Turkey red). Since then many different oils have been sulphated or sulphonated, unsaturated oils as well as oils containing combined hydroxy-groups being used.

The Turkey red oils of the late nineteenth century were usually only partially sulphated and thus contained a proportion of unchanged fatty glycerides. Sulphated oils in which a large part of the glycerides had been hydrolysed to the fatty acids possessed all the faults of the fatty acids themselves, particularly their sensitivity to hard water and to acidic conditions. These defects led to the production (in the 1920–35 period) of so-called "highly sulphonated oils". It should be noted that the continental literature makes little or no distinction between "sulphation" and "sulphonation".

Treatment of unsaturated and/or hydroxylated fatty acids, or their esters, including glycerides, with an excess of sulphuric acid, leads to sulphation of the hydroxyl group and of the double bond. Treatment with more vigorous sulphonation reagents, especially oleum or sulphur trioxide (whether as stabilized liquid or as a gas carried in an inert gas stream), leads to sulphation of hydroxyl groups, sulphonation of the α-carbon atom to the fatty acid or ester group, and to a complex series of reactions with the double bond. Initial reaction with the olefinic group produces a "sultone" group

$$\begin{array}{ccc} -\text{CH} & - & \text{CH} \\ | & & | \\ \text{O} & \!\!-\!\!-\!\! & \text{SO}_2 \end{array}$$

which saponifies to a mixture of ethylenic sulphonates and of hydroxy sulphonates:

$$-\text{CH}{=}\text{CH}{-}\text{CH}_2{-}\underset{\underset{\text{SO}_3\text{Na}}{|}}{\text{CH}}{-}\text{CH}_2{-}$$

$$-\underset{\underset{\text{OH}}{|}}{\text{CH}}{-}\underset{\underset{\text{SO}_3\text{Na}}{|}}{\text{CH}}{-}\text{CH}_2{-} \qquad -\underset{\underset{\text{OH}}{|}}{\text{CH}}{-}\text{CH}_2{-}\underset{\underset{\text{SO}_3\text{Na}}{|}}{\text{CH}}{-}\text{CH}_2{-}$$

the initial α sultone readily rearranging to the β, γ, and δ sultones, all of which hydrolyse partly to the olefinic sulphonic acid and partly to the hydroxy sulphonic acid salt. Excess of sulphur trioxide converts the α sultone to the corresponding carbyl sulphate

$$\begin{array}{ccc} -\text{CH} & - & \text{CH} & - & \text{CH}_2- \\ | & & | \\ \text{O} & & \text{SO}_2 \\ | & & | \\ \text{O}_2\text{S} & \!\!-\!\!-\!\! & \text{O} \end{array}$$

which saponifies to the sulphatosulphonate and then to the hydroxy sulphonate since the sulphonate group activates the hydrolysis of the neighbouring sulphato group. It is hardly surprising that the detailed compositions of the highly sulphonated oils have not yet been fully worked out. A wide variety of means of blocking the carboxyl group, e.g., by formation of the methyl, ethyl, butyl, cyclobutyl, cyclohexyl, and other esters, or by amidification, provides an extensive range of variation in the highly sulphated oils which are widely used in the textile industry because of their excellent wetting-out properties.

Fatty sulphonic acids of this kind were found by the American Twitchell, late in the nineteenth century, to be greatly superior to sulphuric acid as catalysts for the hydrolysis of glycerides. It later proved simpler to condense the fatty acid with benzene, toluene, or naphthalene, and to sulphonate the aromatic ring. The most effective Twitchell reagent is the product from oleic acid and naphthalene.

Treatment of petroleum oils with sulphuric acid as a refining step yielded as by-products complex mixtures of paraffinic and naphthenic sulphonates of dark colour commonly classified as "mahogany acids" or "green acids". The mahogany acids are oil-soluble but the green acids are water-soluble. The neutralized acids are used as emulsifying aids in cutting oils and textile lubricants and in engine oils as sludge dispersants.

Shortage of fats in Germany during the 1914–18 war led to a search for wetting agents and detergents alternative to soap. Naturally, the economic conditions of the time ensured that the search would be made in the field of coal-tar derivatives and the first commercial products were the alkyl naphthalene sulphonates, marketed under the trade name Nekal. Naphthalene, isopropanol and sulphuric acid were heated together. The ratio of isopropanol to naphthalene controls the proportions of mono-, di-, and tri-isopropyl naphthalene sulphonates produced. Of the three, the di-isopropyl product is considerably the more active. The orientation of the groups has not been established: several position isomers are certainly present. The butyl homologues have better detergent properties. With further increase in chain length the most favourable balance of properties shifts from the dialkyl naphthalene sulphonates to the monoalkyl derivatives, excellent products being obtained up to monodecyl naphthalene sulphonate. In the

1920s the German chemical industry pursued a vigorous research programme on synthetic detergents and rapidly produced most of the types now known.

Fatty Alcohol Sulphates

Although sodium cetyl sulphate had first been prepared by Dumas in 1836, commercial interest in the salts of the sulphuric acid hemi-esters was awakened only in the late 1920s. These substances are readily made by reacting the aliphatic alcohol with a considerable excess of sulphuric acid. The lower the temperature of the reaction mixture the cleaner the product. At elevated temperatures various side-reactions may occur leading to the production of an appreciable proportion (up to 25%) of "unsulphated alcohol" which contains variable proportions of dialkyl ether, olefin, sulphone and dialkyl sulphate and little or no aliphatic alcohol. If the reaction is carried out in a suitable solvent at not more than 20°C, products of high quality may be easily produced. More vigorous sulphating agents such as oleum, chlorosulphonic acid, or sulphur trioxide may be used in little more than stoichiometric amounts. In order to secure the maximum yield it is essential that neutralization should be carried out as soon as the sulphation reaction is complete, as quickly as possible with efficient removal of the heat of neutralization, the product of sulphation being run into dilute alkali with vigorous stirring, further alkali being added concurrently to maintain an excess alkalinity throughout the process. Although stable in alkaline solution, the alkyl sulphates hydrolyse with extreme ease at any pH less than 7.

The immediate reaction product is highly soluble in petroleum ether. The course of the reaction is probably best indicated by the equations:

$$R \cdot OH + S_2O_6 + HO \cdot R \rightleftharpoons R \cdot O - \underset{\underset{O}{\|}}{\overset{\overset{O}{\|}}{S}} - O - \underset{\underset{O}{\|}}{\overset{\overset{O}{\|}}{S}} - O \cdot R + H_2O$$

$$R \cdot O - \underset{\underset{O}{\|}}{\overset{\overset{O}{\|}}{S}} - O - \underset{\underset{O}{\|}}{\overset{\overset{O}{\|}}{S}} - O \cdot R + 2NaOH = 2R \cdot OSO_3Na + H_2O$$

Oleyl alcohol, which contains a double bond about the middle of the hydrocarbon chain, may be sulphated with little or no attack on the double bond if the reaction be carried out with the sulphur trioxide adduct of pyridine or dioxane, or in solvents such as ether, dioxane, pyridine, or other solvents basic towards anhydrous sulphuric acid.

Sodium chlorosulphonate is also a useful sulphating reagent. It cannot be made in the wet way but is easily prepared by triturating dry sodium chloride with chlorosulphonic acid, hydrochloric acid being evolved (the reaction must be carried out in a well-ventilated hood, and rubber gloves must be worn).

$$NaCl + HO \cdot SO_2 \cdot Cl \rightarrow NaO \cdot SO_2Cl + HCl$$
$$NaO \cdot SO_2 \cdot Cl + HO \cdot R \rightarrow NaSO_3 \cdot OR + HCl$$

Increased solubility for a given hydrocarbon chain length is obtained if the fatty alcohol is condensed with ethylene oxide prior to sulphation:

$$R \cdot OH + CH_2 \overset{O}{-\!\!\triangle\!\!-} CH_2 = R \cdot O \cdot CH_2 \cdot CH_2 \cdot OH$$

$$R \cdot O \cdot CH_2 \cdot CH_2 \cdot OH + CH_2 \overset{O}{-\!\!\triangle\!\!-} CH_2 = R \cdot OCH_2 \cdot CH_2 \cdot O \cdot CH_2 \cdot CH_2 \cdot OH, \text{ etc.}$$

According to a patent issued to Procter and Gamble, as the length of the molecule is increased in this way the tendency of the tenside to irritate the skin is lessened. Propylene oxide may be used instead of ethylene oxide. Its effect on the temperature of solution is less than that of ethylene oxide, but its effect on irritancy is proportional to the number of carbon atoms, so that for reducing irritancy two propylene oxide units are the equivalent of three ethylene oxide units. The patent claims were supported by extensive animal tests.

The 1930s saw the disclosure by patents of methods of manufacture of the vast majority of the types of tenside now known.

Secondary Alkyl Sulphates

High-temperature cracking of paraffin waxes or petrolatum yields a mixture of olefins which are separated into different distillation cuts. Fractions having a boiling range of about 150–300°C and consisting largely of mono-olefins are reacted with sulphuric acid to yield products consisting mainly (about 80%) of secondary alcohol sulphates, some secondary sulphonates (about 10%), and a proportion of "non-detergent oily matter" (NDOM) of characteristic odour (about 10%). These are generally referred to as "secondary alcohol sulphates". The sulpho group is attached at random along the chain, and the chain lengths cover the range C_8–C_{18} ("broad cut") or C_{10}–C_{18} ("narrow cut"). The products are therefore highly complex mixtures soluble to concentrated solutions at ambient temperatures. By suitable mixtures of cations, as, for example, sodium and potassium, solutions at 15% tenside concentration may be made stable to indefinite storage at 0°C. Because of the high proportion of material with CH_2 chains of moderate length on either side of the carbon atom carrying the sulpho group, these products are outstanding as wetting agents.

To some observers, oxidation of the NDOM by treatment with a little hypochlorite changes the odour from its distinctive "oily" character to one of a perfumery character of intense sweetness, yet to other observers this change is imperceptible.

Fatty Sulphonates

Alkane sulphonates with randomized positions for the sulphonic acid group were produced by the "Reed" reaction (named after the inventor) in which petroleum hydrocarbons of the kerosene or white oil grade (in the United States) or Fischer–Tropsch synthetic hydrocarbons (in Germany) were treated simultaneously with chlorine and sulphur dioxide in the presence of ultraviolet light as catalyst. Sulphuryl chloride could be used in place of the mixture of gases.

$$RH + SO_2 + Cl_2 \rightarrow R \cdot SO_2Cl + HCl$$

Generally, the chlorosulphonyl group may enter anywhere along the chain, but if fatty acids are used in place of the hydrocarbon, the first chlorosulphonyl group tends to go to the β-carbon atom. Polysulphonates are readily produced by this reaction. Like the more familiar secondary alcohol sulphates, these "random" sulphonates are highly soluble at room temperature.

Terminal sulphonates were originally produced by sulphurization of α-olefins followed by oxidation of the thiol or thio compounds produced with, preferably, 70% nitric acid. More recently, cheaper and more satisfactory routes to alkane-1-sulphonates have been opened up by the direct addition of sodium bisulphite to the double bond in presence of a peroxy compound as a catalyst.[†] This process is still under active development. The sodium 1-sulphonates are of lower solubility than the corresponding sulphates.

Straight-chain paraffins are now commercially separated from branched-chain hydrocarbons in white oils and paraffin waxes by selective formation of the urea adducts (characteristic of straight-chain compounds only). It is an instructive experiment to mix 100 g of urea thoroughly with 30 g of dodecyl alcohol at room temperature and initiate the reaction by addition of a few cubic centimetres of water or methanol. The recrystallization of the urea in a less dense form than the normal with absorption of the long-chain alcohol into long channels of molecular diameter is accompanied by a considerable expansion in volume and evolution of heat.

α-olefins may be prepared from the recovered straight-chain hydrocarbons. These may be sulphated with sulphuric acid to yield the secondary 2-sulphates of the parent paraffin, or sulphonated with sulphur trioxide to yield a mixture of unsaturated 1-sulphonate (due to migration of the double bond) and hydroxy sulphonate (due to partial hydrolysis on neutralization of 2-sulphato-1-sulphonate formed by repeated action of sulphur trioxide). These products are commonly referred to as alkene-1-sulphonates to distinguish them from the alkane-1-sulphonates. Pressure to produce biodegradable products

[†] Initial patent disclosures in the early 1930s.

for domestic detergents stimulates much current interest in the alkene sulphonates.

As already mentioned, sulphochlorination of fatty acids by the Reed reaction yields mainly the β-sulphonate of the fatty acid. On the other hand, sulphonation of fatty acids with sulphur trioxide yields the α-sulphonates. These are most effective in use under mildly acid conditions in which the carboxyl group is un-ionized. The disodium salts largely retain the sensitivity to hard water of the ordinary soaps. This can be circumvented by conversion to the methyl ester or the amide, but these improved products are more expensive. Curiously, addition of free mineral acid to the mixture of α-sulphonic fatty acid and methanol is strongly disadvantageous to methylation. Esterification proceeds smoothly and cleanly in the absence of free mineral acid. In spite of continued effort reflected in numerous patents filed from 1928 onwards and the ready availability of the raw materials, little commercial exploitation of the α-sulpho fatty acids has taken place. Unsaturation in the initial oils must first be reduced to less than 1% by hydrogenation. This restriction is less severe if the methyl esters are sulphonated (the upper limit of unsaturated fatty acid esters is about 10%), but the higher cost makes the product less commercially attractive.

Alkylbenzene Sulphonates

Just as economic conditions in Germany, particularly the existence of a powerful chemical industry based on coal, led to the development of short-chain alkyl naphthalene sulphonates, so in the United States the existence of a powerful petrochemicals industry naturally led to the commercial development of long-chain alkylbenzene sulphonates, already disclosed by the German workers. Catalytic alkylation of benzene was an already well-established process, and the attachment of a benzene ring to the long-chain hydrocarbon meant easy sulphonation. By far the greater proportion of the annual production of detergents of approximately 0.5 million tons in the United States and 0.15 million tons in the United Kingdom is of alkylbenzene sulphonates.

In the early stages of this development, kerosenes of molecular weights averaging about C_{12}–C_{14} were monochlorinated and the monochloroalkanes were used to alkylate benzene with the aid of Friedel–Crafts catalysts. These "keryl" benzene "alkylates" were then sulphonated with concentrated sulphuric acid (4–10 moles per mole of alkylate), oleum (1.0–1.5 moles of "free" sulphur trioxide per mole of alkylate) or sulphur trioxide as vapour in air (1.0–1.5 moles per mole of alkylate). Since the former two processes involve the disposal of large volumes of spent acid (in the first process too contaminated to be worth recovery), sulphonation by sulphur trioxide carried in a stream of air has largely displaced them.

Although the keryl benzene alkylates had excellent foaming and derging properties, they fell into disfavour since spray-dried detergent powders compounded with them caked badly on storage in the packet. They were displaced by sodium tetrapropylene benzene sulphonate in which the alkyl chain was made by polymerization of propylene and selection by distillation of an average tetrapropylene cut. Ideally, tetrapropylene benzene (dodecyl benzene) would contain four isopropyl units joined together:

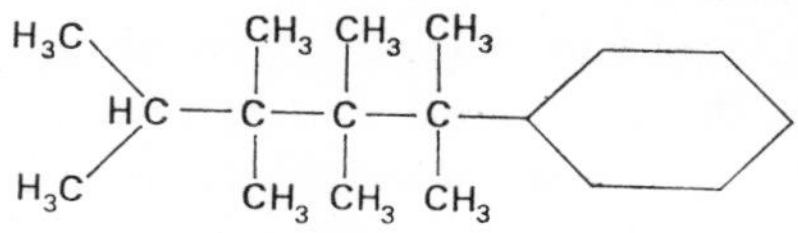

Some tri- and pentapropylene benzenes are also present in the commercial products. In practice, some rearrangement occurs during alkylation of benzene with the Friedel–Crafts catalyst, and the final product is a mixture of isomers. The degree of branching remains high, however, and the sulphonate, passing into sewers from domestic drains, is resistant to breakdown by micro-organisms. It has therefore passed out of the water-purification plants and, particularly in passing down flumes under turbulent conditions, or in passing over weirs, has, possibly in conjunction with other components of the effluent, given rise to the much publicized foaming of effluent channels and streams.

Advances in the technology of powders have made it possible to prevent caking in powders made from sulphonates of the so-called "straight-chain" alkylbenzenes which closely resemble the original keryl benzenes

the benzene ring being attached at random along the straight paraffin chain. Again, owing to partial isomerization during alkylation, some branching of the alkyl chain may be introduced. Mass spectrography permits characterization of alkylbenzenes in terms of the mean number of CH_3 groups per chain. The nearer this is to 1 the more readily and completely biodegradable is the alkane sulphonate. These "soft" alkylbenzene sulphonates are now the preferred type.

The sulphur trioxide may be evaporated into a stream of air either from stabilized liquid sulphur trioxide or from heated oleum. The spent acid is returned to the acid supplier to be re-converted into oleum.

Reaction of sulphur trioxide with alkylbenzene is extremely rapid and evolves a large amount of heat. Hot sulphur trioxide is a strong oxidizing agent. It is essential, therefore, that the sulphur trioxide–air mixture be mixed into the alkylate with intense agitation in order to prevent the formation of temporary local "hot spots". Frequently, a small "heel" of sulphuric acid is added to the reactor in batch processing.

In a continuous process, alkane and a "heel" of 100% sulphuric acid (to 10% by weight of the alkane used) are passed by metering pumps into the first of a series of several sulphonator vessels. The sulphuric acid heel reduces the viscosity of the mixture (which rises rapidly as sulphonation nears completion) and thus assists in the necessary very rapid distribution of the sulphur trioxide into the mass with resulting improvement in the quality of the product. This is less important with the more recent straighter-chain alkylbenzenes which are more fluid than the highly branched tetrapropylene benzene now going out of use.

Sulphur trioxide is generated from 65% oleum heated in a horizontal still to 250°C by stripping with an air stream (pre-dried in a sulphuric acid scrubbing tower). The rate of production of sulphur trioxide is controlled by controlling the in-flow of oleum by means of a rotary feed valve. The sulphur trioxide–air mixture (1 : 2 v/v) is drawn into the sulphonation vessel by a high-speed impeller stirrer running at up to 1400 rev/min (omitted from Fig. 2.1 for the sake of clarity) assisted

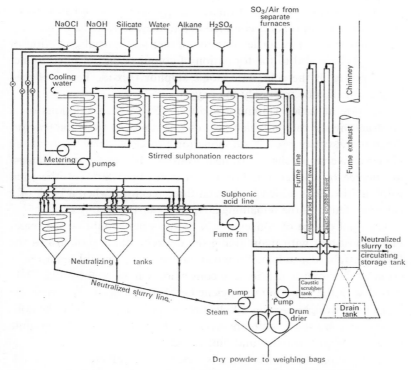

FIG. 2.1. Continuous sulphur trioxide–air sulphonation of alkyl benzene.

by the action of an exhaust pump in the fume line. A separate sulphur trioxide generator is used for each of the sulphonation reactors. These reactors are at decreasing heights above floor level so that the reactant liquid may flow serially through them by gravity.

The sulphur trioxide and alkane react with great rapidity exothermally,

The heat of reaction is removed by the system of cooling coils through which water is passed so as to maintain the temperature below 60°C. As conversion approaches completion and the viscosity rises, oxidative side-reactions may occur leading to rupture and partial oxidation of the alkyl chain and to the formation of colour bodies of intense tinctorial power in the acid state, some sulphur trioxide being reduced to sulphur dioxide to supply the oxygen.

The rate at which sulphur trioxide is fed to each sulphonator is governed by the rate at which heat can be removed from the reactor. Since the viscosity rises as sulphonation proceeds, particularly at the higher conversions, the rate of heat transfer falls rapidly with increasing degree of conversion. For this reason about 80% of the total sulphur trioxide is added in the first three reactors (when a bank of 5 is being used) and the remaining 20% in the last two stages.

The final degree of conversion in the product should be the maximum consistent with maintaining the desired colour standard in the final neutralized product. Provided sufficiently powerful agitation is given to the mixture the limiting conversion will be in the range 97.4–98.1% with the older tetrapropylene benzene and up to 99% with the more fluid straighter-chain alkylbenzenes. Attempts to force the conversion beyond these limits involves the use of considerable amounts of excess sulphur trioxide and much oxidation and charring.

At start-up, each sulphonator is charged with alkane and any sulphuric acid heel required, and sulphonation proceeds batchwise until samples show the conversion levels attained in the steady state on continuous operation. The various feed-valves are then set at the values needed for continuous operation, and a smooth change from batch to continuous is effected. It is unnecessary to empty the plant before short shut-down periods such as week-ends, but if shut-down is to be more prolonged, the plant should be changed back to "batch-run" conditions, the contents of each sulphonator brought to the desired final level of conversion, and the batches transferred to the neutralizing plant and neutralized. The sulphonators may then be washed out and drained.

A system of five half-ton sulphonation reactors, each with 7.6 m² of cooling coil surface, is adequate for a production rate of 2 tons of sulphonic acid per hour.

After leaving the sulphonator the exhaust gases are passed through two scrubbing towers. The first collects entrained acid and the second scrubs the gases with a weak caustic soda solution, the scrubbed effluent gases passing to atmosphere through a tall steel stack.

The sulphonic acid issuing from the last sulphonator passes to any of three neutralizing vessels by way of a common pipeline. These are tanks fitted with cooling coils and stirrers, and may be operated either continuously or batchwise. In the continuous process, 100° Twaddell caustic soda (about 47% w/w) and sulphonic acid flow continuously into the neutralizer. The caustic soda is fed through a metering pump and automatically controlled by a pH controller which monitors continuously the pH of the neutralized paste, issuing from the neutralizer and passing to the storage tank. A small amount of sodium hypochlorite solution may be metered in at the same time in order to bleach the paste to the desired final colour. For some end purposes it is convenient to add small amounts of sodium silicate also at this stage. The concentration of the final slurry is controlled by addition of water.

Since the paste has a strong tendency to separate into curds floating in a serum, the storage tanks are provided with means for continuous circulation by an external pump and suitable pipe-work. After the bleached paste has circulated for 24 hr, the residual chlorine will have fallen to about 0.005% of available chlorine. Since this may bleach dyes added in order to tint products prepared with the paste, the residual chlorine is removed by addition of an anti-chlor such as sodium thiosulphate, sodium sulphite, sodium perborate, etc.

Medialans and Lamepons

Sarcosine (*N*-methyl glycine; $CH_3 \cdot NH \cdot CH_2 \cdot COOH$) is readily and cheaply made on the large scale from methylamine, formaldehyde, and hydrogen cyanide. Condensation of the imino-group with a fatty acid radical leads to the medialans.

$$R \cdot COOH + HN(CH_3)CH_2 \cdot COOH \rightarrow R \cdot CO \cdot N(CH_3)CH_2 \cdot COOH + \\ + H_2O$$

The fatty acid chloride is preferably used in place of the fatty acid itself. The sodium salts of the medialans are much less hydrolysed in water than the unmodified carboxylates, are fairly resistant to hard water, foam well, and are particularly mild to the skin.

If the mixture of amino acids and lower peptides obtained by alkaline hydrolysis of collagen, glue, or scrap leather is used in place of sarcosine, the resulting mixture of fatty amido carboxylates, the lamepons, is even more resistant to hard water than the medialans, and is also capable of dispersing lime soaps. However, the good colour and odour obtainable in the medialans are sacrificed in the lamepons, which are produced in the form of viscous brown liquids containing about 35% of active ingredient.

For either medialan or lamepon production, the fatty acid chloride is fed slowly into the well-stirred solution of amino-acid sodium salts maintained at 25–30°C, caustic soda being added continuously to maintain the alkalinity of the mixture. A short time at about 65°C may follow to finish the reaction.

Alternatively, the sodium soap and the sodium salt of the amino acid, or amino-acid mixture, may be heated together at 130–220°C, or, presumably, the fatty acid methyl or ethyl ester may be heated at 60–120°C with the sodium salt of the amino acids in the presence of 1% or so of sodium methylate or sodium phenate as catalyst, to effect the condensation.

Igepons

When the terminal group is sulphonate instead of carboxylate the products are the igepons, which historically preceded the medialans and lamepons. The general formula may be written

$$RCO \cdot NH(CH_2)_{0\text{-}4} \cdot SO_3Na.$$

The favoured members carry two —CH_2— groups between the amido and the sulphonate groups. The amido group may be replaced by an ester linkage, the —NH— group being replaced by —CH_2—.

In the amido series, the example with only one —CH_2— group between the amido group and the sulphonate group may be made by the attractively cheap and simple process of condensing the amide of the fatty acid with sodium formaldehyde bisulphite. The two components are thoroughly mixed, preferably with a plasticizer such as stearic acid, and heated.

The ester-linked igepon A is made by condensing fatty acid with isethionic acid, either directly in the presence of excess fatty acid as plasticizer, or indirectly by way of the acid chloride or the reaction of sodium soap with sodium chloroethane sulphonate.

$$R \cdot COOH + HO \cdot CH_2 \cdot CH_2 \cdot SO_3Na \longrightarrow R \cdot COOCH_2 \cdot CH_2 \cdot SO_3Na + H_2O$$

$$R \cdot COCl + HO \cdot CH_2 \cdot CH_2 \cdot SO_3Na \xrightarrow[\text{NaOH}]{} R \cdot COOCH_2 \cdot CH_2 \cdot SO_3Na + NaCl + H_2O$$

$$R \cdot COONa + Cl \cdot CH_2 \cdot CH_2 \cdot SO_3Na \longrightarrow R \cdot COOCH_2 \cdot CH_2 \cdot SO_3Na + NaCl$$

This tenside, the fatty acids consisting of a mixture of tallow and nut oils acids, is the major component of the first commercially successful "non-soap-detergent" toilet bar which has now been on the US market for several years.

Although igepon A is sufficiently stable to hydrolysis for normal purposes, it is readily hydrolysed in hot, strongly alkaline conditions. The amide-linked products are much more stable to hydrolysis. Igepon T, KT, and 702K are, respectively, the oleyl, coconut, and palm-oil fatty acid condensates with methyl taurine:

$$RCOCl + NH(CH_3)CH_2CH_2SO_3Na$$

$$\xrightarrow{\text{NaOH}} RCONH(CH_3)CH_2CH_2SO_3Na + NaCl$$

(the corresponding product made from taurine, in which the nitrogen atom carries a reactive hydrogen atom, are skin sensitizers). These products have been of considerable commercial importance in Germany, whose chemical industry produces methyl taurine at a reasonable cost.

The N-methyl substituent may be replaced by other alkyl groups or by rings, and the N-alkyl group may be long, e.g. $C_{16}H_{33}$, in which case the acyl group will be that of a short-chain fatty acid, e.g. acetic acid, thus

$$CH_3CON(C_{16}H_{33})CH_2CH_2SO_3Na,$$

or the alkylene chain linking the N atom to the sulphonate group may consist of zero to four CH_2 groups, or of a phenyl group. The phenyl group, again, may be substituted, for example with CH_3 and/or OH groups.

Isethionic acid is made by condensing ethylene oxide with sodium bisulphite, the bisulphite presumably reacting in the iso form

$$\underset{CH_2 \cdot CH_2}{\overset{O}{\diagup \diagdown}} + H \cdot \overset{\overset{O}{\|}}{\underset{\underset{O}{\|}}{S}} \cdot ONa \longrightarrow HOCH_2CH_2SO_3Na$$

Methyl taurine is then made by reacting the sodium isethionate with methylamine at 270–290°C and under 200 atm pressure

$$CH_3NH_2 + HOCH_2CH_2SO_3Na \longrightarrow (CH_3)NHCH_2CH_2SO_3Na + H_2O$$

"Reversed" amide sulphonates have been made by reacting a fatty amine with chloroacetyl chloride and subsequent sulphitation:

$$RNH_2 + ClCH_2COCl \longrightarrow RNHCOCH_2Cl$$
$$\xrightarrow{Na_2SO_3} RNHCOCH_2SO_3Na + NaCl$$

The high cost of fatty amines has made this type commercially unattractive. However, the potentially cheap route to fatty amines now available by reaction of α-olefins with hydrogen cyanide and the like may again bring such products into practical consideration. Analogous ester-linked types are the alkyl sulphoacetates $ROOCCH_2SO_3Na$ made by esterification of a fatty alcohol with chloroacetic acid followed by sulphitation:

$$ROH + ClCH_2COOH \longrightarrow ROOCCH_2Cl$$
$$\xrightarrow{Na_2SO_3} ROOCCH_2SO_3Na + NaCl$$

Owing to its high cost, this attractive product is exploited mainly in cosmetics.

Sulphosuccinates and Sulphocarballylates

Sodium bisulphite is readily added across a double bond activated by neighbouring polar groups as in the mono- and di-esters of maleic acid:

$$\begin{array}{ccc} R_1OOC \cdot CH & & R_1OOC \cdot CH_2 \\ \| & + NaHSO_3 \longrightarrow & | \\ R_2OOC \cdot CH & & R_2OOC \cdot CH \cdot SO_3Na \end{array}$$
$$\text{di-alkyl sulphosuccinate}$$

These form the group of tensides introduced under the name "aerosols". The best-known member of the series is aerosol OT, the dioctyl ester in which the octyl radical is 2-ethyl hexyl. The aerosols have powerful wetting action.

Related compounds are made by combining long-chain mono-esters of maleic acid with, for example, glycol or ethanolamine and further condensing these products with ethylene oxide. Addition of bisulphite to the double bond converts these into alkyl sulphosuccinic acid polyalkylene oxides or alkylsulphosuccinic acid amide polyalkylene oxides, which are claimed to be exceptionally non-reactive to the skin and to have little or no enzyme inhibitory action.

Similarly, citric acid triesters may be dehydrated to trialkyl aconiₐates and sulphited with sodium sulphite:

$$C_6H_{13}OOC \cdot CH_2$$
$$|$$
$$C_6H_{13}OOC \cdot C \cdot OH \longrightarrow$$
$$|$$
$$C_6H_{13}OOC \cdot CH_2$$

$$C_6H_{13}OOC \cdot CH$$
$$||$$
$$C_6H_{13}OOC \cdot C$$
$$|$$
$$C_6H_{13}OOC \cdot CH_2$$

$$C_6H_{13}OOC \cdot CH_2$$
$$|$$
$$\xrightarrow{\text{NaHSO}_3} \quad C_6H_{13}OOC \cdot C \cdot SO_3Na$$
$$|$$
$$C_6H_{13}OOC \cdot CH_2$$

to form trialkyl sulphocarballylates. The trihexyl compound is a powerful wetting agent.

With longer alkyl chains, it is preferable to use the monoesters in order to retain solubility, e.g.

$$C_{12}H_{25} \cdot OOC \cdot CH_2 \cdot CH \cdot COOH$$
$$|$$
$$SO_3Na$$

As with the alkyl sulphates, the introduction of a chain of ethyl ether groups between the alkyl and the ester linkages enhances both the aqueous solubility and the mildness to the skin. Alkyl polyether sulphosuccinates have been introduced under the name "Condanol" and are reputed to be notably mild on the skin, e.g.

$$RO(CH_2CH_2O)_n OC \cdot CH(COOH)SO_3Na$$

where n represents an average value.

Polyalcohol Monoalkyl Sulphates

Sulphated "mono" glycerides may readily be prepared by first sulphating glycerin with oleum at about 30°C, adding a triglyceride or mixture of triglycerides and heating until the product is completely soluble in water at a reaction temperature of about 50°C. The product probably contains a small proportion of diacyl glycerin sulphate. In the neutralization stage the pH of the reacting mixture must be held within the range 7.5–10.5, too high a pH permitting hydrolysis of the fatty ester grouping and too low a pH permitting hydrolysis of the sulphate ester group.

Over-production of cane-sugar stimulated attempts to produce sugar-based detergents. The difficulty of reacting sugar with, say, a fatty acid, lay in the diverse solvent requirements of the two components. This was solved by the discovery that dimethylformamide was a solvent for both sugar and for fatty substances. Although the technical problems of manufacturing fatty esters of sugar were brilliantly solved, and the product could subsequently be sulphated to produce an anionic detergent, the cost of the process has worked against the commercial exploitation of such products.

Epichlorohydrin reacts with fatty alcohols to yield chloroglycerin monoethers which on sulphitation yield "glyceryl monoether sulphonates" (3-alkoxy, 2 hydroxy propane sulphonates).

$$R \cdot OH + CH_2 \cdot CH - CH_2Cl \xrightarrow{NaOH} R \cdot O \cdot CH_2 \cdot CH(OH) \cdot CH_2Cl$$

$$\xrightarrow{Na_2SO_3} R \cdot O \cdot CH_2 \cdot CH(OH) \cdot CH_2 \cdot SO_3Na + NaCl$$

These are stable to hydrolysis and are attractive detergent products. An alternative process eliminates the sodium chloride in an intermediate stage

$$R \cdot O \cdot CH_2 \cdot CH \cdot CH_2Cl \xrightarrow{NaOH} R \cdot O \cdot CH \cdot CH_2 + NaCl$$
$$\qquad\qquad OH$$

the epoxide being reacted with bisulphite to yield the final product.

Non-ionic Detergents

Prohibition of thiodiglycol (which reacts with anhydrous hydrochloric acid to form mustard gas) in Germany after 1918 stimulated investigation of glycol condensation products. These had been recognized as linear polymeric materials by Lourenco in 1868, only some 15 years after the first isolation and characterization of ethylene glycol itself. Condensation products of ethylene glycol are readily obtained in the laboratory by the reaction of chlorohydrin, in the presence of caustic alkalis, on compounds containing replaceable hydrogen atoms such as alcohols, phenols, carboxylic acids, aldehydes, ketones, methylene, or ethylene groups activated by the near presence of ketonic or carboxylic ester groups, mercaptans, amines, imines, and amides. For many such products, solution in excess ethylene carbonate

$$(O{=}C{-}O{-}CH_2$$
$$\diagdown O{-}CH_2$$

formed by the action of phosgene on glycol) followed by heating to 95–100°C, and the addition of a molar quantity of anhydrous potassium carbonate, affords a simple laboratory route to clean products. Industrially, ethylene glycol condensation products are prepared by the addition of ethylene oxide to the compound carrying reactive hydrogen atoms at superatmospheric pressure and at moderately elevated temperatures. Ethylene oxide may be made by the action of concentrated alkali lye on chlorohydrin, but this route has been largely displaced industrially by the catalytic oxidation of ethylene with air over a platinum catalyst.

The preparation of ethylene oxide condensation products was vigorously explored by the large German chemical concerns, with the natural extension to include similar epoxides, particularly propylene oxide,

$$CH_3CH \cdot CH_2$$

butylene oxide,

$$CH_3 \cdot CH_2 \cdot \overset{\displaystyle O}{\overset{\diagdown}{CH}} \cdot CH_2$$

and glycide or glycidol, the epoxide derived from glycerin,

$$CH_2(OH) \cdot \overset{\displaystyle O}{\overset{\diagdown}{CH}} \cdot CH_2$$

and the related chlorohydrins. Exploitation of the auto-condensation of these epoxides (in the initial presence of a little water), of their reaction with reactive hydrogen atoms in organic functional groups, and of the esterification or etherification of pre-formed polyglycol, etc., chains (which carry terminal hydroxyl groups) with organic acids or alcohols, led to a spate of patent activity which rapidly disclosed a vast field of compounds, soluble or dispersible in water and of high solubility in fats and oils, exceptionally versatile in the range of their applications, and capable of being "tailor made" for optimum activity in specific uses. These industrial uses include wetting-out, emulsification, cleansing, dye-levelling, fabric softening, dispersing (particularly useful in the preparation of drilling muds), fat-liquoring of leather, machinery lubrication, spinning oils, solvents for organic matter, and many other uses.

The attached hydrophobic groups could be straight or branched, saturated or unsaturated hydrophobic chains, saturated or unsaturated hydrocarbon rings, or heterocyclic rings, alone or combined in the same molecule, non-carbon linking atoms being permissible as were substituents such as hydroxy, carboxy, ester, sulphonic acid, halogen, nitro, amino, or amido atoms or groups.

The condensates were found to be compatible with soaps and other compounds "of a saponaceous character", gums, glues, acids, bases, and salts. A particular advantage, disclosed almost as an aside, in a 1930 patent directed to improved dye levelling agents, was that by adding to "Marseilles soap" from 3% to 15% octadecan-1,12-diol (obtained by hydrogenation of castor oil) condensed with 15 molar parts of ethylene oxide, complete stability to very hard water (45°H French) was attained.

Successive condensation to the hydrocarbon compound containing at least one reactive hydrogen atom, of different epoxides, in any desired order and proportion, was already disclosed in patents of the late twenties and early thirties, as was the quaternization of one or more nitrogen atoms in the molecule to form cationic derivatives, with or without the additional attachment of carboxy, sulphato or sulphonato, phosphato or phosphonato groups to form betaines, sulphobetaines, and phosphobetaines of high surface activity. Similarly, products to which only a few epoxy groups had been attached and which were therefore not water-soluble but, at most, water-dispersible, could be sulphated or phosphated, or reacted with, e.g. sodium chloroacetate, to provide an anionic, water-solubilizing, terminal group. On the other hand, properties could be modified in the other direction by esterification of the terminal hydroxy group by, for example, a fatty acid.

Full appreciation of the great extent of the early work can be gained only by careful reading of the original patents, full reference to which are to be found in the larger and more comprehensive books such as those by Moilliet, Collie, and Black, and by Schwartz and Perry.

Reaction of ammonia with alkylene oxides leads to a mixture of mono-, di-, and trialkylolamines, separable by distillation. Condensation of the mono- and dialkylolamines with fatty acid chlorides, in Germany in 1929, produced a fresh class of non-ionic compounds, the fatty alkylol-amides, largely insoluble in water but valuable as emulsifiers, and as lather improvers in conjunction with water-soluble foaming and washing agents. Many of these are referred to more fully in Chapter 3. Further condensation of the fatty alkylol-amides with alkylene oxides brings them into the water-soluble class of non-ionics.

When the fatty material is an alcohol, the initial chlorohydrin or ethylene oxide reacts at once to produce a small concentration of the monoether of glycol:

$$R \cdot OH + \overset{\displaystyle O}{\overset{\displaystyle \diagup\diagdown}{CH_2 \cdot CH_2}} \longrightarrow R \cdot O \cdot CH_2 \cdot CH_2 \cdot OH$$

The product is therefore also a primary alcohol and will have the same reactivity towards further ethylene oxide as the original fatty alcohol. Thus some monoalkyl diglycol will be produced very early in the course of the reaction, and from this, again, monoalkyl triglycol, and so on. The concentration of the original alcohol will fall at a decreasing rate throughout the reaction. The concentrations of the successive condensation products will rise to a maximum and then fall slowly, as the reaction proceeds. After a given time, a whole series of products will be present, the proportion of each being given by the Poisson distribution series. According to this statistical distribution, if n is the average number of ethylene oxide molecules condensed with one molecule of fatty alcohol, then the relative frequency of occurrence of the various integral condensates are as follows:

No. of ethylene oxide molecules per fatty alcohol	Relative frequency
0	e^{-n}
1	$ne^{-n}/1!$
2	$n^2e^{-n}/2!$
3	$n^3e^{-n}/3!$
4	$n^4e^{-n}/4!$
etc.	etc.

The significance of this is readily shown by calculation of some typical examples (see Table 2.2). For convenience the relative frequency has been multiplied by 100 to give per cent occurrence in the mixture.

It is clear that as the reaction proceeds the distribution curve flattens out and spreads over a continually increasing number of species.

Reaction of ethylene oxide with a carboxylic acid is more rapid than with a primary alcohol, hence the free acid disappears rather more rapidly than is indicated in the following table for the initial reactant. Conversely, the amide group reacts far more slowly than the primary alcohol group and hence considerable amounts of unreacted amide are present even when the "average" number of moles of condensed ethylene oxide has reached ten or even twenty. These variations

TABLE 2.2. PER CENT FREQUENCY OF OCCURRENCE OF INDIVIDUAL SPECIES

Individual Species	$n =$			
	1	3	5	7
$R \cdot OH$	36.8	5.0	0.67	0.09
$R \cdot O(CH_2CH_2O)_1H$	36.8	14.9	3.37	0.64
$R \cdot O(CH_2CH_2O)_2H$	18.4	22.4	8.4	2.23
$R \cdot O(CH_2CH_2O)_3H$	6.13	22.4	14.0	5.21
$R \cdot O(CH_2CH_2O)_4H$	1.53	16.8	17.6	9.12
$R \cdot O(CH_2CH_2O)_5H$	0.31	10.1	17.6	12.8
$R \cdot O(CH_2CH_2O)_6H$	0.05	5.04	14.9	14.9
$R \cdot O(CH_2CH_2O)_7H$		2.26	10.7	14.9
$R \cdot O(CH_2CH_2O)_8H$		0.81	6.7	13.0
$R \cdot O(CH_2CH_2O)_9H$		0.27	3.7	10.1
$R \cdot O(CH_2CH_2O)_{10}H$		0.08	1.9	7.1

in reactivity can be mitigated to some extent by proper choice of pH, temperature, and catalyst, but cannot be altogether eliminated. The initial reactions of an epoxide with an acid amide would be as follows:

$$R \cdot CONH_2 + CH_2 \cdot CH_2 \longrightarrow R \cdot CONH \cdot CH_2CH_2OH$$

$$R \cdot CONH \cdot CH_2CH_2OH + CH_2 \cdot CH_2 \longrightarrow R \cdot CO \cdot N \begin{array}{l} CH_2 \cdot CH_2OH \\ CH_2CH_2OH \end{array}$$

that is, first a monoethanolamide is produced which might react further to produce the diethanolamide; but this is not a practical process. It is much more satisfactory to prepare these substances directly (as described in Chapter 3) and then to condense with ethylene oxide to produce the desired polyether products than to attempt production from fatty amide and ethylene oxide.

The alkyl phenols condense readily with ethylene oxide and have had considerable commercial exploitation, particularly in the field of wool scouring.

The complex mixture of fatty and resin type carboxylic acids known as "tall oil"—a by-product of the sulphate process of recovering cellulose for papermaking from softwood chips—is a cheap source from which polyethenoxy monoesters may easily be manufactured. Lanolin, lanolin alcohols, and beeswax alcohols yield polyethenoxy ethers useful in cosmetics.

Members of the alkylphenol series with only one or two average glycol residues, the unreacted fraction having been removed by distillation, are good lather promoters. The crude materials are useful as emulsifying agents and as lather breakers. Members whose $\bar{n}$ is 8 or more are water-soluble detergents. The sulphated fatty alcohol polyethanol ethers have already been mentioned.

A distinctly separate type of non-ionic product has enjoyed commercial development in the United States. These are known as the "spans" and "tweens". The "spans" are simple esters of the higher fatty acid with hexitol–hexitan mixtures obtained by reduction of sugars to sugar alcohols; addition of ethylene oxide converts "spans" into "tweens". The "spans" are of low water solubility, the "tweens" are readily soluble in water. These relationships have already been displayed in Table 1.5 (p. 56) in terms of their HLB values.

Glycerin undergoes auto-condensation when heated in the presence of a little caustic soda or sodium methylate as catalyst. The degree of polymerization may be followed by following the change in refractive index which accompanies it. Whereas glycerin fatty monoesters are emulsifying tensides, polyglycerin monoesters of higher fatty acids are water-soluble non-ionic detergent tensides: a suitable average degree of polymerization is five. When the refractive index shows that this degree of polymerization has been reached, the caustic alkali is neutralized and a small excess of sulphuric acid is added as esterification catalyst. Fatty acid is added to one-fifth to one-third of that required to form the monoester in order to minimize the formation of di- and higher esters, and heating is continued until esterification is complete. On addition of saturated sodium sulphate the esters separate from the water–polyglycerol electrolyte solution as an upper layer. The lower layer is run off. The unreacted polyglycerol may be recovered by evaporating off the water under reduced pressure. The polyglycerol ester product gives a harsh, open lather.

Cationics

Alkyl halides may be reacted with primary, secondary, or tertiary amines to produce secondary or tertiary amines and quaternary ammonium salts. Usually, it is preferred also to quaternize secondary or tertiary amines produced in a first stage of reaction. The tertiary amino nitrogen atom may be contained in a ring system as in pyridine. Typical compounds of this class are:

> cetyl trimethyl ammonium bromide,
> cetyl dimethyl benzyl ammonium chloride,
> cetyl pyridinium chloride.

The majority of these quaternary, cationic tensides are powerfully germicidal and are extensively used, along with strong alkalis such as sodium metasilicate and caustic soda, in the commercial washing of milk bottles. They are also used in hand "scrubbing" liquids for surgeons.

Many cationics are strongly toxic. Compounders of products intended for widespread domestic use are therefore rightly chary of incorporating a cationic detergent in these products. However, since antiseptics and disinfectants are generally appreciated as toxic, many domestic antiseptics are based on cationics.

The toxicity of cationics appears to be reduced in the presence of ether linkages in one or more of the attached hydrocarbon chains and if the trimethyl groups are replaced by alkylol groups. Such substituents also decrease the tendency to mutual precipitation of anionic and cationic detergents and permit the use of stable salts of cationic-anionic detergent ions which, within particular series, may exhibit a sharp "peak" of foam stability at a particular value of $n+m$ where n and m are the numbers of carbon atoms in the cationic and anionic detergent major alkyl chains. Even within a limitation of total number of carbon atoms to not more than forty nor less than ten, the number of combinations of four groups (which may be straight or branched chain, alicyclic, aromatic, or heterocyclic, with nought to many of various substituents and with nought to, say, five intermediate ether, thioether or other linking groups) is so great that it must be many years before all the possibilities of cationic surface active chemicals have been explored.

Cationics have been among the more expensive tenside products; the potentially cheap route to fatty amines, mentioned above, by addition of hydrogen cyanide to α-olefins followed by hydrogenation, may eventually lead to rather lower cost materials.

Amide-linked quaternary compounds such as N-methylene stearamide pyridinium chloride obtained by reaction of chloromethyl stearamide and similar compounds with tertiary amines are unstable to heat. As cationic compounds they are readily adsorbed from aqueous solution on to cellulosic fabrics. On heating after drying, the compounds decompose at the tertiary nitrogen atom, the tertiary amine is volatilized, and the methylene stearamide radical either dimerizes or forms a chemical bond with the cellulose, to form a water-repellant surface layer which is very fast to washing. This forms the basis of the ICI's Velan process of water-proofing fabrics.

Other uses of cationic tensides are as fabric-softening agents, for which purpose they are applied in a "rinse" wash, strongly adsorbed fatty substances imparting a notably soft "handle" to fabrics, as wash-fastness-imparting agents for direct cotton dyestuffs, and as levelling agents for vat dyestuffs.

Ampholytes

Ampholytes are those compounds which may be either basic or acidic, according to the pH of the system, or both acidic and basic simultaneously. If both the acidic and basic functions are weak, as in the alkylamino carboxylic acids, the former holds as, for example,

$$\left.\begin{array}{l} \overset{+}{C_nH_{2n+1}\cdot NH_2\cdot(CH_2)_mCOO'} \\ C_nH_{2n+1}\cdot NH\cdot(CH_2)_m\cdot COOH \end{array}\right\} \text{ "neutral"}$$

$$\begin{array}{ll} \overset{+}{C_nH_{2n+1}\cdot NH_2(CH_2)_m\cdot COOH} + X' & \text{acid pHs,} \\ C_nH_{2n+1}\cdot NH(CH_2)_m\cdot COO' + M^+ & \text{alkaline pHs} \end{array}$$

If both functions are strong then the zwitterion[†] structure exists over a wide pH range spanning the neutral point, as in

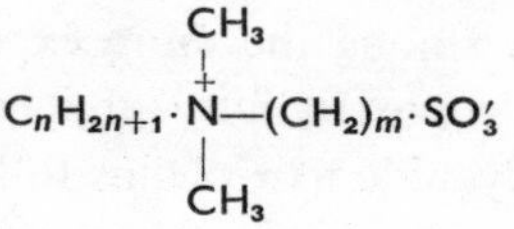

$$C_nH_{2n+1}\cdot \overset{CH_3}{\underset{CH_3}{\overset{|}{\underset{|}{N^+}}}}-(CH_2)_m\cdot SO_3'$$

or in

$$C_nH_{2n+1}-\overset{+}{N}\underset{\diagdown CH_2\cdot COO'}{\overset{\diagup CH_2COOH}{-CH_2\cdot COOH}}$$

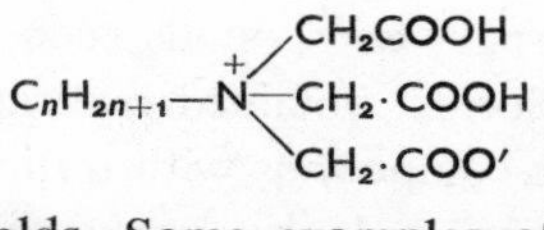

and the latter case holds. Some examples of this class of tenside exhibit strong bactericidal powers with lower mammalian toxicity than shown by many quaternaries.

Miranols are made by the reaction

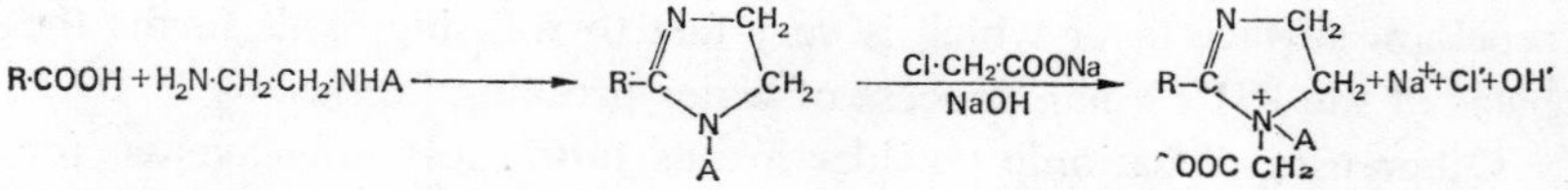

where A is lower alkyl, hydroxyethylene or aminoethylene. When A is aminoethylene, $-CH_2\cdot CH_2\cdot NH_2$, further stages of condensation with sodium chloroacetate are possible, to such products as

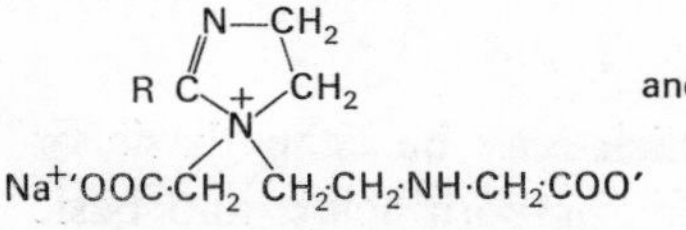

and

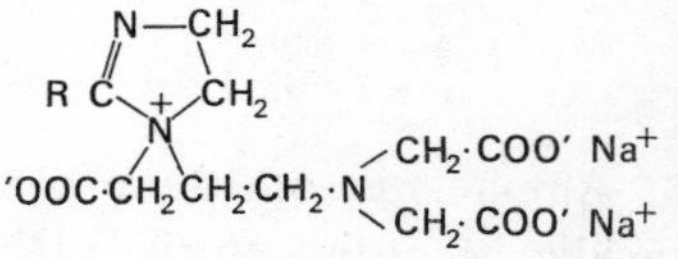

These are both tensides and sequestrants.

The miranols are said to be mild on the skin, germicidal, and compatible with both cationics and anionics.

_____

[†] Simultaneously cationic and anionic.

Reading List

RICHARDSON, K. N. Automation in the soap industry, *Soap, Perfumery and Cosmetics*, Jan. 1965, p. 43; Feb. 1965, p. 137.

ZILSKE, H. Modern soap boiling, *Perfumery and Essential Oil Record* **56,** 314 (1965).

HILDITCH, T. P. and WILLIAMS, P. N. *The Chemical Constitution of Natural Fats*, 4th edn., Chapman & Hall, 1964.

SCHWARTZ, A. M., and PERRY, J. W. *Surface Active Agents*, vol. 1, 1949, vol. 2, 1958, Interscience.

MOILLIET, J. L., COLLIE, B. and BLACK, W. *Surface Activity*, 2nd edn., Spon, 1961.

CHAPTER 3

Ancillary Surface Active Chemicals

Lather Promoters

The characteristic "soapy" lather is copious and dense. Pure non-soap tensides usually give tenuous "open" foams. If caustic alkali be added to soap solutions to suppress the hydrolysis, the soap foam also becomes tenuous and open. In the section on adsorption in Chapter 1 it was shown that, although the hydrolytic fatty acid was present in only small amount in the bulk solution, it constituted about 50% of the surface film. Addition of excess fatty acid to soap raises its proportion in the surface layer still further and tends to make the lather even more "creamy".

It is characteristic of long-chain polar, non-ionizing (or non-ionized) compounds that they are very strongly adsorbed. Packing of such molecules into a close-packed film is not opposed by powerful electrical repulsive forces, indeed, it may often be promoted by hydrogen-bonding as well as by dipole–dipole attractive forces. With insoluble films on water, changes in area available to the film molecules in the "gaseous film" or "expanded" film regions involve only small changes in the surface pressure: changes of area available per molecule in the "condensed film" region involve large changes in pressure. Under equilibrium conditions, surface films on solutions of long-chain compounds bear little relation to surface films of insoluble compounds since the composition of the film is invariant with change in area. But under the non-equilibrium conditions of mechanical "shock" or rapid variations of available surface, the relatively slow diffusion of molecules from the surface when the area is expanded sharply, causes the behaviour of the soluble film to approximate momentarily to that of the insoluble film: the surface concentration of molecules falls and the surface tension rises sharply. This means that the Gibbs elasticity (Chapter 1, section on foams, p. 45) will be large and the foam will be stable.

It is not surprising, therefore, that the closeness and stability of foams of non-hydrolysing ionic tensides are improved by the addition of a proportion of polar non-ionizing compounds.

Quite early in the development of non-soap detergents it was observed that the crude products of sulphation of fatty alcohols foamed much better than the purified materials. The crude material contained a proportion, of the order of 10%, of unsulphated fatty matter which was assumed to be unreacted alcohol. A patent claim was made for the improvement of the lathering power of fatty alcohol sodium sulphates by the deliberate addition of that fatty alcohol from which the product had been made. This addition has the expected effect. Most of the extracted unsulphated fatty matter, however, was not unchanged fatty alcohol but by-products substantially free from hydroxyl groups and is no longer sulphatable. It contains some sulphur and is probably a mixture of fatty ether, fatty diester of sulphuric acid, and olefin, the proportions of these by-products varying with the reaction conditions used.

Fatty amides and alkanolamides were also early recognized as lather improvers.

Fatty amides are manufactured by reacting ammonia with fatty acid at a high temperature, the apparatus first being purged with nitrogen. Preferably an autoclave is used and ammonia pumped in from time to time until the pressure ceases to drop. At atmospheric pressure a reflux column is required to return fatty acids carried over with the stream of ammonia. The temperature should be maintained between 170° and 190°C. At lower temperatures the dehydration of the ammonium soap to amide is sluggish, while at higher temperatures further dehydration to nitrile occurs. At 210°C the nitrile is formed in major quantity.

Monoalkanolamides (ethanolamides and isopropanolamides are those commonly manufactured) are readily prepared by heating the fatty acid and ethanolamine (isopropanolamine) together at about 170°C. The diethanolamides may also be made in this way but the method in this instance has two disadvantages: (i) a large excess of the amine must be used, and (ii) the major product is the ester rather than the amide. After completion of the reaction at high temperature, the temperature must be lowered to about 110°C, when slow conversion from ester to amide takes place. A much more favourable method for the dialkanolamides is by the reaction of a lower fatty ester (methyl or ethyl esters for preference but glycerides may be used) with the dialkanolamide at temperatures from about 60° to about 110°C in the presence of a small proportion of a strongly alkaline catalyst, e.g. $\frac{1}{2}\%$ of sodium methoxide or of sodium metal. This reaction is substantially stoichiometric (95–98% yield of amide) and no excess of dialkanolamine is required. The reactants are not mutually miscible and must be well stirred. If the stirring be stopped from time to time, the two phases rapidly separate and for some time the relative volumes appear to remain unchanged. After a time which decreases with increasing temperature, the biphasic mixture rapidly becomes homogeneous, this stage of the reaction taking only a few seconds. By adding the reactants with catalyst to pre-formed acyl diethanolamide, the process can be made continuous.

The monoesters and monoethers of polyols such as glycol, glycerin, and pentaerythritol, also afford useful lather improvers. Separation of the monoesters from di- and tri-esters adds to their cost. The mono-ethers are easily prepared by reacting glycol chlorohydrin or α-chloro-hydrin with fatty alcohols, or with alkyl phenols. Commercially, products such as glycerin monolaurate or glycerin monostearate are available, "technical" grades containing about 30% "mono" and "high mono" grades about 70%. Their main outlet is not in the detergent field but in the bakery trades where they are valued as emulsifiers.

Since the essential feature of the molecular structure of lather improvers is the presence of a strongly polar group, preferably capable of hydrogen bonding, at the end of a sufficiently large non-polar group, the variety of potential lather improvers is considerable. In addition to the types of compound already mentioned, long-chain ureides, sulphoxides, and amine oxides include efficient lather improvers. Long-chain 1-halides and 1-nitro compounds have not provided efficient examples.

Many non-soap tensides have the useful property of dispersing the scum formed by soap in hard water. It has been found that combinations of ionizing tensides with the non-ionizing tensides used as lather improvers are commonly much more effective than either class of tenside alone.

Hydrotropes

The effects on each other of different solutes present together in an aqueous solution are many and varied. The "salting out" of various solutes such as soaps, dyestuffs, alcohols, on addition of electrolytes, is a common and well-known phenomenon. The opposite effect of "salting in" is rather less common and, in the normal sense, is usually of minor magnitude.

Many organic substances are more soluble in mixtures of alcohol and water, or of acetone and water, than in either solvent alone. This "mixed solvent" effect is explained by the altered "solubility parameter" of the mixture, discussed in Chapter 1. Frequently, the solubility increases sharply over a narrow range of compositions of the mixed solvent.

There exists a large group of solutes with which gross increases in solubility in water of organic substances, of moderate to extremely low solubility in pure water, may be obtained. The phenomenon has been known for over a hundred years and was commercially exploited at least by the early 1870s in the preparation "Lysol"—an aqueous solution of soap and cresol used as a disinfectant in which it was known that substantial quantities of paraffinic hydrocarbons could be dissolved to a clear solution which remained clear on dilution. These compositions were extensively investigated by Engler and Dieckhoff in the years 1887 to 1892. The solubilizing action of urea and thiourea on alkaloids, proteins, cyclohexane, aniline, cinnamic aldehyde, and other substances was known as early as 1900, and only 7 years later the principle was enunciated that if a substance is soluble in each of two partially miscible liquids, it increases their mutual solubility. The solubilizing activity towards condensed ring aromatic hydrocarbons of purines in low concentration is extraordinary.

Neuberg carried out a detailed investigation of these and similar instances. He greatly extended the range of known examples and in 1916 distinguished this behaviour as a distinct phenomenon which he defined as "hydrotropy" (Greek *hydros* = water; *tropos* = a turning; to turn towards water). The substances exerting this effect he called "hydrotropes". Constitutive influences were investigated. For instance: in the *o*-, *m*-, and *p*-substituted benzoic acids the hydrotropic power decreases in the order *o* > *m* > *p* (nil) when the substituent is hydrophilic, e.g. hydroxyl, but remains unchanged when the substituent is hydrophobic, e.g. methyl.

Alkyl sulphates of moderate chain length and benzene sulphonates are as hydrotropic as the benzoates: hydrotropy is therefore not confined to soluble carboxylates. Starch, proteins, alkaloids, and inorganic salts are solubilized by hydrotropes as readily as hydrocarbons, polar oils, and long-chain surface active agents: hydrotropy is therefore not simply due to the hydrocarbon portion of the molecule exerting its own solubility in the concentrated aqueous solutions of the hydrotropes. Not only are insoluble proteins solubilized but they and the soluble proteins are rendered heat stable and non-coagulable in the presence of hydrotropes.

With increasing molecular weight of the hydrotrope, the hydrotropic power rises to a maximum at about C_7 in the aliphatic series and at C_8–C_{10} in the aromatic series. The general structure is thus similar to that of the detergents, hydrophobic and hydrophilic portions of the molecule being spatially distinct, although attached to each other. Several dyestuff intermediates are strongly hydrotropic, e.g. 1-hydroxy-2naphthoates, 2 hydroxy-naphthalene-1-sulphonates. Among the xylene sulphonates, the *ortho* is rather better than the *meta* which is rather better than the *para* compound. *m*-Xylene sulphonate, however, is greatly preferred owing to its much greater solubility. Ethyl benzene sulphonate, cymene, cumene, and tetralin sulphonates are also effective hydrotropes.

The relative solubility increase of the solubilized solute, defined by $R = (S_H - S)/S$, where S_H is the solubility in the hydrotrope solution and S the solubility in pure water, depends on the concentration of the hydrotrope according to the empirical relationship

$$R = R_1 c^n \qquad\qquad (3.1)$$

or $\qquad\qquad \log R = \text{constant} + n \log c.$

The value of n is independent of the hydrotrope. This simple rule will not apply to micelle-forming solutes.

Although hydrotropes are widely used in the formulation of liquid products containing high concentrations of detergents (15–40%) and have a secondary effect in reducing the viscosity of these compositions, the phenomenon of hydrotropy is of much broader applicability. For example, by increasing the rate of solution and the limiting solubility of esters, hydrotropes greatly increase their rate of hydrolysis. This could be of importance in the detoxification of mustard gas. Other organic reactions taking place in aqueous media may be speeded-up wherever their rates are limited by the limited solubility of one or more of the reactants.

It has sometimes been noticed that crystalline complexes may be obtained with hydrotrope plus solubilized compound, and it has been suggested that hydrotropy is an expression of complex formation between hydrotrope and other solute in the solution.

However, it is by no means universally true that there must be a 1 : 1 correspondence between the formation of solid crystalline compounds and compound formation in aqueous solution. On the one hand, many compounds are capable of crystallizing into lattices containing unfilled spaces or channels large enough to accommodate other molecules. Even the inert gases may be obtained in crystalline "complexes" in this way, in which instances there can be no question of chemical combination of however loose a character. The only requirement is good geometrical fit. Such complexes can exist only as crystals, and fall apart completely when dissolved in water. On the other hand, many complexes exist in solution which have to be dissociated in order that the solute may be crystallized out. The numerous examples of anhydrous crystalline salts, the dissolved ions of which are hydrated, such as sodium chloride, are well-known instances. Of course, should a complex formed in solution be of a shape capable of forming a crystal with a greater decrease in free energy than is involved in the formation of crystals of the pure solute, a crystalline complex will be readily obtained.

The possibility that compound formation may occur in the solution and afford a mechanism for solubilization must not, however, be rejected out of hand. The empirical relationship given above [eqn. (3.1)] is readily deducible from the mass action law. Let the formation of the complex be represented by

$$B + nH \rightleftharpoons BH_n. \tag{3.2}$$

Then

$$K = \frac{[BH_n]}{[B][H]^n}. \tag{3.3}$$

B, the concentration of the solute as simple molecules, may be taken as the saturation concentration in the water, and may therefore be absorbed into the reaction constant, thus:

$$K' = (BH_n)/(H)^n. \tag{3.4}$$

BH_n, the concentration of the complex, is the total concentration of the solute S_H minus the amount dissolved in the absence of hydrotrope S, i.e.

$$(BH_n) = (S_H - S).$$

H, the concentration of free hydrotrope, is the total concentration of hydrotrope C minus n times (BH_n), i.e. $c - n(S_H - S)$. Substituting these values into eqn. (3.4) we have

$$K' = (S_H - S)/[c - n(S_H - S)]^n. \tag{3.5}$$

Dividing both sides by S,

$$k = \frac{K'}{S} = \left(\frac{S_H - S}{S}\right)\bigg/[c - n(S_H - S)]^n. \qquad (3.6)$$

When the ratio of c to S is large and n is of the order unity, $n(S_H - S)$ is small compared with c, and, to a fair approximation,

$$k = (S_H - S)/Sc^n \qquad (3.7)$$

or

$$R = R_1 c^n, \qquad (3.1)$$

where R is the ratio $(S_H - S)/S$.

It must, however, be borne in mind, that the fact that data with certain combinations of solute and hydrotrope can be fitted by the equation does not prove that complex formation in solution is the mechanism of action of the hydrotrope.

The limitations of this derivation will be evident from the values of n given in Table 3.1.

TABLE 3.1. VALUES OF n FOR VARIOUS HYDROTROPE–SOLUTE COMBINATIONS

Solute	Ben-zene	Hex-ane	Hep-tane	Cyclo-hex-ane	Benzo-ic acid	Phthal-ic acid	Benz-pyrene
Hydrotrope	—	—	—	—	—	—	—
Sodium butyrate	4.45	4.8	—	8.1	—	—	—
Sodium caproate	2.8	4.7	4.2	3.95	—	—	—
Sodium heptylate	2.6	3.45	3.4	3.9	—	—	—
Sodium caprate	1.65	2.3	1.65	2.1	—	—	—
Sodium hexahydro-benzoate	2.3	4.3	5.0	5.9	—	—	—
Sodium benzene sulphonate	—	—	—	—	1.70	1.11	—
Sodium toluene sulphonate	—	—	—	—	1.80	1.22	—
Sodium ethylbenzene sulphonate	—	—	—	—	2.04	1.16	—
Caffeine	—	—	—	—	—	—	1.74

Marked deviations in additivity of optical rotations have been observed in special instances in which both hydrotrope and solute were optically active. This could be due to complex formation, or due to the marked change in the character of the solvent. On the other hand, measurements of rates of diffusion of benzoic acid and of brucine, dissolved in aqueous sodium benzene sulphonate, showed, according to Freundlich, that the solutes were undoubtedly dissolved as completely free molecules. It must be concluded that although complex formation in solution between hydrotrope and solute may sometimes occur, it is not an essential prerequisite for hydrotropy to be displayed.

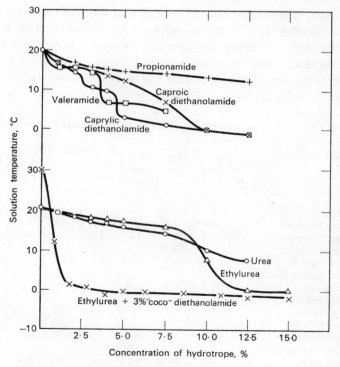

FIG. 3.1. Temperature of solution to 10% of sodium tetrapropylene benzene sulphonate in the presence of hydrotropes.

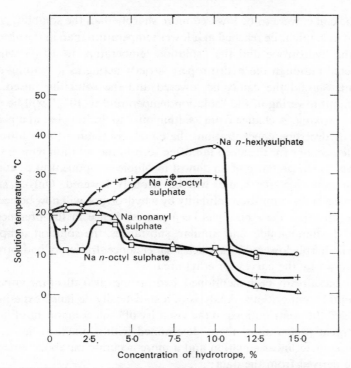

FIG. 3.2. Temperature of solution to 10% of sodium tetrapropylene benzene sulphonate in the presence of hydrotropes.

An example of the complex interactions between long-chain surface active solutes, hydrotropes, and inorganic electrolytes has already been given in Fig. 1.4 of Chapter 1 (p. 19). The effects of a selected number of hydrotropes on the temperature of solution to 10% of salt-free sodium tetrapropylene benzene sulphonate are now shown in Figs. 3.1 and 3.2.

It will be recalled that the solubility curves of detergents are usually characterized by a sharp upward sweep at the temperature at which the saturation solubility to single molecules reaches the critical micelle concentration formation (c.m.c.). Hydrotropes may lower both the c.m.c. and the solubility, especially at low hydrotrope concentrations.

8*

If the c.m.c. were to be lowered more sharply than the solubility, then the c.m.c. might be reached at a lower temperature than in the absence of the hydrotrope and the "solution temperature to 10%" will be lowered although the hydrotrope is strictly acting as a "salting-out" agent. Should the c.m.c. be lowered and the solubility raised, the resultant lowering of the "solution temperature to 10%" will be even more striking. A change from "salting out" to "salting in" at a particular hydrotrope concentration, the c.m.c. continuing to be lowered, could account for the steep fall often seen in the solution temperature curve. In the presence of an almost insoluble component like "coco"-diethanolamide, the c.m.c. will be strongly lowered. Only a small increase in the saturation solubility by a hydrotrope will now be needed to ensure that the c.m.c. has been passed, hence in the presence of coco-diethanolamide and similar substances, the solution temperature is sharply lowered by a much smaller concentration of hydrotrope, as shown by the curves for ethyl urea.

As has already been mentioned, hydrotropes also affect the viscosity of liquid compositions. Analysis of a statistically designed experiment enabled the contributions to the viscosity of each component of liquid detergent preparations to be determined within practical limits. An approximate linear equation and a more accurate parabolic equation were derived from the data:

$$\log \eta = \text{constant} + a[TD] + b[H_1] + c[H_2] + \ldots, \tag{3.8}$$

$$\log \eta = \text{constant} + a[TD] + b[H_1] + c[H_1][TD] + d[TD]^2 + \ldots. \tag{3.9}$$

The coefficients a, b, c, d, $\ldots$, are the contributions to $\log \eta$ of each 1% of the component in the mixture, $[TD]$ is the total concentration of tenside, $[H_1]$ that of hydrotrope 1, and so on. The viscosity η is in centipoise. For example: by eqn. (3.8), with the constant $= -4.5$, an "active" (detergent + lather improver) concentration of 20%, plus 5% sodium xylene sulphonate, 10% alcohol and a, b, and c coefficients of $+0.35$, -0.14, and -0.09 respectively, the viscosity would be given by the antilog of -4.5, $+7.0$, -0.7, -0.9, i.e. antilog 0.9, hence a viscosity of about 8 centipoise. The coefficients are to be determined for each blend of actives.

For the particular systems studied, the coefficient a in eqn. (3.9) was positive but the square term coefficient d was negative. These vary with the type of detergent and lather improver and are capable of being separated into individual coefficients for each of these. The coefficients b, c, etc., for the hydrotropes were negative but the cross-terms c were positive. Thus the effect of these hydrotropes on the viscosity decreased rectilinearly with increasing detergent concentration. Within the range of concentration stated above, the effect of sodium xylene sulphonate and ethanol remained negative, but for urea it dropped to zero and then became small and positive as the total detergent concentration increased through about 22%. The determination of the coefficients for any system of interest is a straightforward statistical operation.

Reading List

McBain, J. W. and Hutchinson, E., *Solubilisation*, Academic Press, 1955.
Winsor, P. A., *Solvent Properties of Amphiphilic Compounds*, Butterworth, 1954.

CHAPTER 4

The Manufacture of Adhesives

Animal Glues

Hide and Bone Glues

Glue and gelatin are derived from the collagen of animal hides and the ossein of bones. Essentially, the process of extraction involves the unwinding of the triple helix of the insoluble protein to produce three separated strands of protein, now soluble in warm water. Some hydrolytic degradation also occurs. Treatment may be by acidic or alkaline solutions or simply by hot water. The products obtained by the acidic and alkaline processes respectively differ sufficiently to lead to different isoelectric points, pH 8.5 in acid-processed gelatins and pH 4.5 in alkali-processed gelatins and glues. Glue extracted from bones by hotwater treatment resembles alkali-processed glue in having an isoelectric point at about pH 4.5. Acid processes are comparatively little used.

The most advanced processes for the extraction of bone glue are those of Cheyen and of Vyner. In the Cheyen process, crushed bones are subjected to high-frequency mechanical impulses generated by a rotor driven at 2500 rev/min and bearing forty-five beaters. The rotor operates above a semi-circular screen 24 in. long with $\frac{1}{4}$ in. and 1 in. openings. The bones are fed between the two as an aqueous slurry. The fat cells are ruptured by the impulses and the fat is washed off the bones to collect as a creamy layer on the surface of the water. This process yields fuel economy, lower processing costs, and an improvement in quality in comparison with degreasing by solvent extraction. In the Vyner process rapid treatment at low temperatures is relied on. The fat is removed after pretreatment and crushing of the bones and is separated from the water by conventional means. The Vyner process also provides tallow of greatly improved quality compared with solvent extraction processes, leaving the collagen relatively undamaged.

Glue is then extracted from the degreased and washed bones by alternate steam and water treatments. In the conventional process steam at $5–30$ lb/in² is followed by hot-water leaching. The aqueous liquid and the bones move in counter-current through a number of vessels. Guidotte claims that with sufficient comminution of the bones, steaming is unnecessary.

In addition to the breakdown of the ossein to glue, a mucopolysaccharide–protein complex is extracted. The hydrolysate of this complex has been shown to contain five sugars and to have an amino-acid composition corresponding to that of the blood-serum proteins rather than that of ossein or collagen. The mucoids are likely to be represented in bone glue by their degradation products, and it may be that they play a significant part in the special properties of bone glue.

Hoofs and horns are keratinous rather than collagenous, but contain collagenous piths extractable on treatment with steam or hot water for short periods.

Hide glues are prepared from washed skins after removal of non-glue proteins by treatment with a suspension of calcium hydroxide in water. After washing, light acidification, and re-washing, the hides are "cooked" in a series of tanks, the hides and liquors moving in counter-current, much as in the preparation of bone glues. The higher grades of hide glue constitute the gelatins used in the food, pharmaceutical, and photographic industries.

The $3–7\%$ liquors from the cookers are evaporated to $15–55\%$ total solids content for sale as "tanker" or "jelly" glues. The jelly may be further dried to form cake glues which may be ground to provide powdered glues. Preservatives such as sodium pentachlorophenate (to about 1%) may be added with advantage.

Fish Glues

Non-oily skins of large fish, continuously available in high tonnage, are used in the manufacture of fish glue. Cod skins, in particular, meet the requirements. A ton of skins yields 400–450 lb of liquid fish glue.

To prevent bacterial decomposition, the skins are stored in a salted and dried condition. After a wash treatment to remove the salt, the skins are cooked in a series of vessels in counter-current. The liquors, drawn off at 5–7%, are concentrated to about 50% solids.

Blood Glues

The fibrin which is responsible for the clotting of blood must be removed quickly from fresh blood intended for the manufacture of a stable adhesive. Acid precipitation or mechanical agitation is used to remove the fibrin. The residual blood, containing haemoglobin and serum proteins, is stabilized by addition of preservative. Blood proteins denature and become insoluble on heating above about 70°C. The fibrin-free blood must be dried, therefore, under carefully controlled conditions to retain water solubility. Usually, spray-drying or vacuum drying at temperatures below 70°C is used to produce the most soluble dry blood glues which retain 80–95% solubility in cold water. By careful control of the temperature, dried bloods of lower solubility (in the range 15–80% of the blood soluble in cold water) may be obtained as required. Bloods which have been overheated and rendered insoluble are of little use as adhesives.

Casein Glues

Acidification of skim milk to pH 4.5 precipitates the protein casein as a curd. Three pounds of washed-and-dried curd are obtained from 100 lb of milk. Bacterial fermentation of the milk sugar to lactic acid, or addition of mineral acid, may be used for the acidification. When mineral acid is used, lactose may be recovered in addition to casein. Neutralization of some of the acidic groups in the protein re-solubilizes casein to form glues. Organic or inorganic, fixed or volatile alkalis may be used. Care must be taken to ensure that the pH does not rise too high in order to avoid degradation by hydrolysis to low-molecular peptides. This reaction proceeds rapidly at pH levels of 10 or more. A volatile alkali may evaporate as the joint ages, leaving insoluble casein in the glue line. More effective water-resistant casein glues are obtained with formulations containing lime as part of the alkali. In order to secure a reasonable pot-life, a soluble sodium salt and lime in excess of that required to convert the sodium salt to caustic soda are dry-mixed with dry, powdered casein. When the mixed powders are dispersed in water, rapid reactions convert the sodium salt to caustic soda which forms sodium caseinate, and a slow reaction finally converts the sodium caseinate to insoluble calcium caseinate. The made-up glue dispersion must, of course, be used before the slow reaction has proceeded so far as to make the glue useless. Suitable sodium salts are sodium silicate (particularly those grades with ratios of $SiO_2 : Na_2O$ of $3 : 1$ to $3.5 : 1$), trisodium phosphate, soda ash, and sodium fluoride. Variations in the amounts and proportions of the sodium salts used and the degree of excess of the lime, permit wide variation of the viscosity, pot-life, and water resistance of the glue. Control of these properties to meet particular requirements tests the skill of the formulator.

Water-resistant glues may be made without lime if oxides of other di- or poly-valent metals, or their salts, are used instead. Alternatively, water-resistance may be achieved by use of formalin, formaldehyde donors, or carbon disulphide.

Other ingredients of a complete casein-glue formulation include oils, thiourea, thickeners, preservatives, and fillers. Addition of a small amount of glyceride oil prolongs the pot-life of the glue. Urea, sodium hexametaphosphate, sodium sulphite, sugar, or dicyandiamide may be added to thin the glue; formalin, soluble calcium salts, alums, or sodium sulphate may be added to thicken it. Organic meals such as wood flour and the like are useful fillers, restraining penetration of the glue into the members of the joint and aiding gap filling. Chlorinated phenols, soluble fluorides, borax, and mercurials such as phenyl mercuric acetate, are all effective preservatives.

Vegetable Glues

Soyabean Glues

Soyabean meal, from which the greater part of the oil content of the beans has been extracted by hydraulic pressure or by solvent extraction (at moderate temperatures [$\Rightarrow$ 70°C] to avoid reducing the solubility of the protein) is a source of protein valuable as an adhesive. The meal is ground to flour with a specific surface area of $3-6 \times 10^3$ cm²/g and contains some 50% protein, 35% carbohydrates, 1% residual oil, mineral matter, and water. In their natural state the protein molecules are tightly coiled, and in this condition have poor adhesive properties. To prepare soyabean flour as an adhesive it is necessary to slurry it with water and to raise its pH to 11 or more. At this pH level the protein molecules uncoil irreversibly with development of their adhesive capacity. They also become subject to hydrolysis. Soyabean glues are therefore prepared "on site" and have a useful pot-life of only a few hours.

Increased pot-life and an improved water resistance of the dried glue-line may also be obtained with soyabean glue by the addition of soluble calcium salts, polyvalent metal oxides or their salts, formaldehyde donors, etc. Polymer latex emulsions may be added and give excellent water resistance. The carbohydrate fraction of the flour, in the alkaline preparation, is more than an inert filler, the starches becoming solubilized and contributing appreciably to the "tack" of the glue.

Considerable care must be exercised in dispersing the flour in alkaline aqueous media in order to avoid lumpiness and frothing. It is advisable to disperse the material in water first, adding the alkali only when a smooth dispersion has been obtained. Efficient well-powered mixers are essential equipment. A defoamer must be added at an early stage: it may be blended into the dry powder after grinding to flour. Pine oil to about 3% on the flour is effective.

As with all animal and vegetable products, preservatives are necessary to prevent attack by moulds, fungi, and bacteria. In addition to preservatives already mentioned, copper 8-quinolinate, copper naphthenate, and tributyltin oxide are also useful.

Starches and Dextrins

Starches consist of mixtures of a highly branched glucose polymer, amylopectin, and an unbranched polymer, amylose. The starches characteristic of different plant species differ widely in the proportions of amylose and amylopectin present, and probably also in the structural details of the amylopectin fraction. In the United States, corn (maize) starch, and in Europe, potato starch, is most generally used. A variety of maize (waxy maize) which produces an entirely amylopectin starch, has been produced by genetic selection. A variety producing a purely amylose starch has not yet been developed, but may be expected to become available in time. Starch grains from different plant species vary in size from about 5 μ to about 90 μ in diameter, and are also distinguished by characteristic forms and markings. The grains swell rapidly in hot water within a narrow temperature range lying within the region 60–77°C, the precise value, however, being another species characteristic. On continued heating in water, the grains finally burst, and the starch is dispersed to a colloidal sol.

The viscosity of a starch sol depends on many factors: the species of plant from which the starch was obtained, the previous history of the sample, the concentration of electrolytes and their nature, the pH, and the technique of preparation of the sol. The amylose fraction readily recrystallizes from the sol at room temperature, a process resulting in "set-back" or gelling of the starch sol and the staling of bread and cakes. Dried films of amylopectin are brittle or powdery; those of amylose are tough. Addition of borates outstandingly increases tack and stability.

Partial hydrolysis with acids reduces the viscosity and leads to the "thin-boiling" or "soluble" starches. For the less fluid products the starch is processed in acid slurries, neutralized, recovered, and dried.

For the more fluid products the starch is swollen with acidified water and dried. Enzymic hydrolysis may also be used.

Dextrins also are made by heating acidified starch in the dry state. The nature of the changes that occur in the dextrinization of starch is still obscure. It is believed that hydrolysis of the original starch is accompanied by re-polymerization, the final product containing many more and much shorter branches than the initial material. Variation of initial moisture and acidity of the starch, the temperature at which it is heated, and the time for which it is heated, permit the production of a wide range of dextrins. Specific differences between starches of different botanical origin can also influence the character of the dextrins. "White" dextrins still retain some of the pasting property of the original starch. The deep amber dextrins known as British gums show no trace of pasting but form transparent solutions of high viscosity. "Canary" dextrins fall between these extremes. The distinction between successive classes is, however, quite arbitrary, products of any desired degree of conversion being obtainable.

Modified Starches

Starch in aqueous dispersion is readily oxidized by sodium hypochlorite. Quite a low degree of oxidation stabilizes the viscous starch sol and greatly improves its film-forming properties. Oxidized starch is particularly valued for the surface sizing of paper.

Starch may also be methylated, hydroxymethylated, carboxymethylated, etc., in the same manner as cellulose (see next section). The products are useful as thickeners but find little outlet as adhesives, the dextrins being so much cheaper.

Cellulosics

Cellulose is an unbranched polymer of glucose and is insoluble both in water and in organic solvents although soluble in concentrated aqueous solutions of some salts. Cellulose esters, on the other hand, both inorganic as in the cellulose nitrates and organic as in cellulose acetate or propionate, are soluble in organic media, whilst ethers such as methyl cellulose, hydroxyethyl cellulose, and sodium carboxymethyl cellulose are soluble in water. However, ethyl cellulose is also soluble in organic solvents.

Nitration is by a mixture of nitric and sulphuric acids; acetylation by means of acetic anhydride in presence of pyridine or sodium acetate. The ethers are made by the action of alkyl halides such as methyl chloride or sodium chloroacetate, on cellulose suspended in a strongly alkaline medium (70% isopropanol in water) at about 70°C. Hydroxyethylcellulose may alternatively be made by reacting ethylene oxide with soda cellulose.

The preparation of adhesives from cellulose derivatives is a matter of formulation. Cellulose nitrate, for example, is dissolved in a suitable solvent mixture such as acetone, ethanol, and amyl acetate in the proportions 30 : 14 : 56. To 100 parts by weight of solvent, 15–20 parts of nitrocellulose and 6–8 parts of camphor (as plasticizer) are added. Organic esters may be preferred as being less flammable. Thirty parts by weight of cellulose acetate–butyrate are dissolved in 100 parts of solvent consisting of toluene, ethanol, and nitropropane-1 in the proportions 75 : 20 : 5 with addition of 3 parts of tricresyl phosphate as plasticizer. Synthetic resins such as polyvinyl acetate may also be added. Ethyl cellulose formulations yield adhesives tough at low temperatures; water, heat, and alkali resistant; and free from discolouration in sunlight. Twenty-five parts of ethyl cellulose in a 75 : 25 toluene–ethanol solvent with 1 part of octyl phenol as plasticizer forms a suitable basis, addition of synthetic resins again being optional.

The water-soluble cellulosics afford a grease-resistant glue-line. Aqueous dispersions of methyl cellulose have the property, unique among gums, of gelling on heating. In preparing these dispersions it is essential first to swell the material with boiling water. On cooling the mixture, the methyl cellulose disperses to a colloidal sol of high viscosity at concentrations around 2%. Sodium carboxymethyl cellulose is readily soluble in both hot and cold water. Methyl cellulose and carboxy methyl cellulose are widely used in modern wallpaper adhesives.

Competition has led to the development of a wide range of cellulose ethers, prominent commercial products being ethyl cellulose, hydroxyethyl cellulose, and higher alkyl and hydroxyalkyl celluloses. Variations in the mean molecular weight of the cellulose and in the degree of substitution make possible an extended series of products showing fine gradations in properties. In testing blends of methyl and carboxymethyl cellulose to secure a desired effect, over eighty different products were examined: of all the combinations possible only two satisfied the conditions.

The variations possible in solubility characteristics of short-chain cellulose ethers implies that the range of products includes examples of outstanding emulsifying agents. Not only are they adsorbed at the interface so as to prevent coagulation of the emulsion by providing a strongly water-attracting outer layer for the emulsion drops, but by virtue of their high molecular weight and high viscosity–plasticity at a high concentration (such as the adsorbed layer must be) they afford a strong mechanical resistance to coalescence of the drops. Ethyl celluloses may be chosen for either water-in-oil or for oil-in-water emulsions. In the same way, by adsorption at solid–liquid interfaces they exert highly effective dispersing action, the wide range from water-soluble to oil-soluble products ensuring that by suitable choice of ethyl cellulose, the finely divided solid may be dispersed in any desired medium.

Cellulose caprylate is a thermoplastic material which replaces Canada balsam as a lens cement with advantage since it does not discolour with age and it remains permanently thermoplastic, enabling the lens to be dis-assembled, cleaned, and re-assembled if need be.

Rubber

Natural rubber occurs as an emulsion ("latex") in the saps of several plants. The rubber derived from the South American tree *Hevea brasiliensis* is by far the most widely known and the most important. It is a polymer of *cis*-isoprene. The droplets of polymerized unsaturated hydrocarbon in the latex are stabilized by an adsorbed film of protein.

The latex of commerce is usually further stabilized by the addition of a little ammonia and increased in concentration from about 40% rubber solids, as drawn from the tree, to about 60%. The latex is easily coagulated. The coagulum, prepared in sheet form, may be "smoked" as a form of preservation to yield crepe rubber.

Crepe rubber is dispersible in hydrocarbon solvents yielding a rubber "solution" having marked adhesive properties. However, raw natural rubber does not afford a particularly satisfactory adhesive, and it is usual to add vulcanizers ("curing agents"), tackifiers, fillers, plasticizers, sequestrants for heavy metals, and antioxidants.

"Vulcanization" or "curing" implies the development of intermolecular cross-linking. This decreases solubility, ultimate elongation, and deformability; increases the elastic modulus, the rate of elastic recovery, and resistance to creep. Sulphur is the most common vulcanizer. "Tack" is enhanced by addition of rosin or rosin derivatives at the cost of some loss of final bond strength. Some fillers, particularly carbon black, may appreciably strengthen the rubber ("reinforcement"); others are either inert or weaken the bond. Since rubber is a highly unsaturated hydrocarbon it is readily degraded by oxidation. Numerous antioxidants have been patented which considerably extend the "induction period" during which little oxidation occurs. These are generally phenols, polyphenols, or aromatic amines. They apparently act largely as breakers of free-radical chains. Oxidation being strongly catalysed by even minute traces of transition metals, particularly copper and manganese, heavy metal sequestrants are added.

Crepe rubber is prepared for solution by a process known as "mastication" in which it is subjected to severe mechanical working. It is squeezed between heavy steel rollers run at different speeds and is intensely sheared. The heat developed is partially dissipated by water-cooling the rollers internally. During mastication chemical bonds are broken by the intense mechanical effort with production of free radicals and reduction of the average molecular weight which lies in the range 0.25×10^6 to 2.5×10^6 in the raw rubber. The free radicals produced are chemically highly reactive. They may either recombine, form side chains, or combine with other chemicals present, e.g. with oxygen, to form peroxides.

Rubber reclaimed by severe mastication from old tyres and other scrap is not only cheaper than raw rubber but has several advantages, particularly in higher adhesiveness and faster setting, when used in adhesive formulations.

Rubber solution adhesives are made with up to 25% rubber solids in toluene or other suitable solvent, the more fluid examples containing 10–15%. The masticated rubber is treated with solvent in a heavy-duty mixer. The solvent is added in successive small portions, each portion being thoroughly worked into the mass before the rest is added, until at least half the solvent has been incorporated, when the remainder may be added more quickly. The other additives may be blended-in either during mastication, concurrently with solvent addition, or after solvent addition. If blended with the rubber during mastication, the tackifier should be added, as nearly as possible, at the end of the operation.

Mastics, putties, and sealers are high in solids content, with a high proportion of filler. Asphalt or bitumen is often included. Tackifiers are used in much lower proportions than in the rubber solution cements. Mastics are made in the same way as the solution cements.

Rubber latex may be used directly as an adhesive. The films produced after the water has evaporated are tougher and stronger than films from solvent cements since the rubber molecules have not been broken down by mastication. Improvement by addition of tackifiers, etc., is still desirable. The additives are blended into the latex in emulsion form in order to avoid coagulation of the latex. Subsequent heat treatment still further improves the bonding characteristics of the latex. Reclaim rubber may also be dispersed in aqueous media to latex-like products. The reclaim rubber is softened by mastication and the desired fillers, softeners, sulphur, etc., added towards the end of this process. The plasticized compound mass is transferred to a heavy-duty mixer and tackifier blended in slowly as needed. Water is added, at first in small quantities. When about 10% of the total water has been added, the chosen dispersing agent is added and blended in.

More water is added, little by little, to about 30% of the rubber taken, and mixing is continued until the water-in-rubber emulsion inverts to a rubber-in-water dispersion. Heating the mass, or addition of some hot water as this stage is approached, hastens the inversion. Finally, the dispersion is diluted to the required concentration (about 40% solids) by addition of the remainder of the water.

Synthetic Glues

Rubber-like Polymers

The art and science of polymerization form a distinct technology which will not be discussed here. The products of this technology, however, now afford a comprehensive range of adhesives.

The rubber-like polymers—elastomers—comprise homopolymers of dienes and copolymers of dienes with unsaturated nitriles, carboxylic acids, esters, etc. The dienes commonly used are isoprene, chloroisoprene, and butadiene. The other components commonly used for these products are acrylonitrile, acrylic acid, methacrylic acid, methyl acrylate, ethyl acrylate, and styrene. Although the elastomers are confined to those copolymers in which the diene is the major component, the number of different types of polymer possible within this limitation remains large.

Very broadly, the synthetic rubbers may be classified as:

(i) butyl and polybutylene rubbers;
(ii) nitrile rubbers;
(iii) styrene-butadiene rubbers;
(iv) carboxylic rubbers;
(v) polychloroprene rubbers.

Polybutylene rubbers are homopolymers of butadiene and are almost completely saturated paraffins in character. They are therefore inherently much more stable to oxidation than the highly unsaturated natural rubber. A quite small proportion, no more than 0.5% is necessary, of isoprene copolymerized with butadiene increases the unsaturation sufficiently to confer on the product sufficient cross-linking capacity to permit profound changes in viscosity, ultimate elongation, and strength by use of the ordinary vulcanizing agents. Nitrile rubbers containing upwards of 40% of acrylonitrile are highly resistant to petrols and exhibit enhanced adhesiveness compared with polybutylene. Styrene–butadiene copolymers are particularly abrasion-resistant. Synthetic rubbers with free carboxyl groups are exceptionally good for bonding metals to metals. Polychloroprene is both resistant to petrol and to abrasion.

The polymers may be produced directly in solvent systems, requiring only the appropriate additives for compounding to solvent adhesives, or as latexes which may be coagulated and the coagulum milled and blended, or the latex may be mixed with the desired additives, such as already described for natural rubber. Basic oxides such as magnesium oxide and zinc oxide must always be used with polychloroprene to absorb any acids formed by hydrolysis or by the action of light, and thus inhibit the processes of degradation.

Formaldehyde Condensates

In the presence of dilute acid or alkali as catalyst, phenol and formalin react at ambient temperatures as follows:

If an excess of formalin be used, further substitution occurs:

The elements of water are readily eliminated from the hydroxy methylene group and the benzene nucleus of another molecule, particularly on heating:

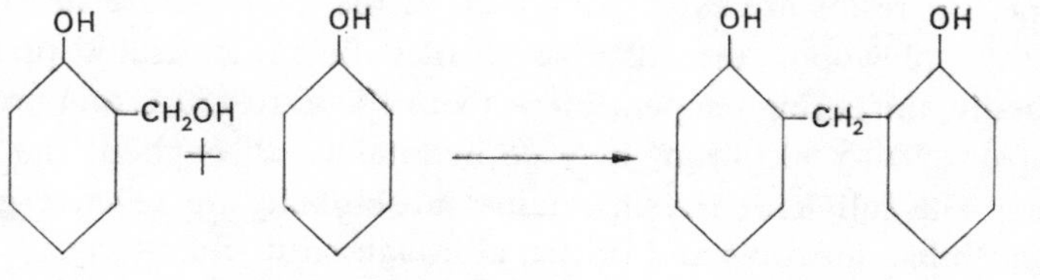

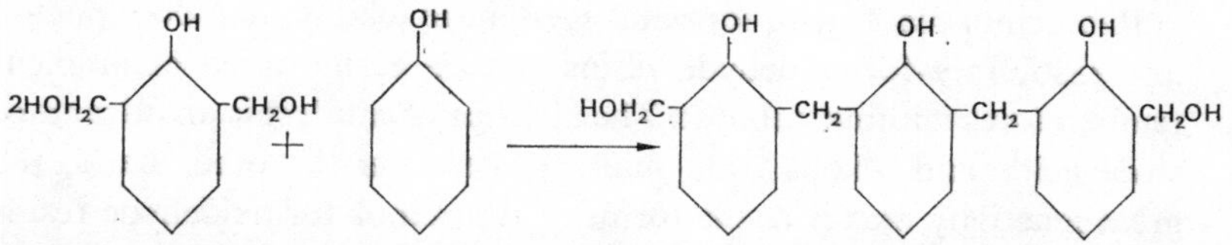

The phenol-terminated chains, formed readily in acid solutions with up to 25% molar excess of phenol, are called "novolacs". They are soluble in water, fusible, and stable. The hydroxymethylene terminated chains, formed readily in alkaline solution with up to 50% molar excess formalin, are called "resoles". They are soluble in water, readily soluble in alkaline media, and are unstable, particularly on warming. By continued condensation between molecules they build up solid three-dimensional structures, the phenol–formaldehyde (PF) resins:

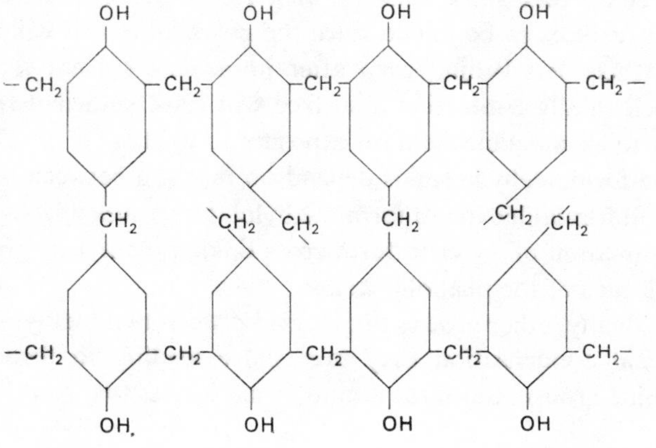

The resoles must be used fairly soon after manufacture. The novo-lacs may be stored indefinitely and "cured" to form resins by addition of formaldehyde-releasing substances such as hexamethylene tetra-mine. PF resins are extensively used in the manufacture of exterior-quality plywood, the alkaline resoles being generally preferred. Usually, the curing temperature is from 120°C to 150°C and pressures of from 50 to 500 lb/in² may be maintained throughout the curing time. The fully-cured resin retains substantially no reactive groups, is insoluble, infusible and brittle, although hard and strong.

Resorcinol reacts with formalin even more readily than does phenol and resorcinol–formaldehyde resins can be easily cured at ambient temperatures without catalyst. The absence of catalyst is an advantage since acids and alkalis stain many woods. On the other hand, the great reactivity makes resole forms of resorcinol–formaldehyde resins impossible to use. The working material is a syrup obtained by condens-ing 1 mole of resorcinol with 0.60–0.65 mole of formaldehyde. The syrup is stabilized and fluidized by addition of ethanol to a level of about 65% non-volatile content.

A formaldehyde-releasing agent such as paraformaldehyde is used as the curing agent and is conveniently added ready compounded with any filler or reinforcing agent which it may be desired to use. Wood-flour, nut-shell flour, or bark flour may be used. The final mixture, at about 25°C, may have a pot-life of from about 1.5 to about 4 hr.

Resorcinol-resin glues must be kept rigorously off surfaces other than the surfaces to be joined since the resorcinol itself will stain the wood. Traces left behind even after immediate solvent removal of resin accidentally contacting a surface will leave sufficient residue to give rise to an indelible stain on exposure to light.

Amine formaldehyde resins depend on reaction between the amino group and formaldehyde to form methylol derivatives which condense with elimination of water to form cross-linking methylene groups just as in the case of the phenolic resins.

Additionally, ether groups may form between two methylol groups, or internal condensation may take place leading to formation of azomethine groups. Since the amino group carries two reactive hydrogen atoms, each amino group may react with two molecules of formaldehyde. The major amino compounds used in making adhesives are urea and melamine:

$$O=C\begin{array}{c}NH_2\\NH_2\end{array}\ +2HCHO\ \longrightarrow\ O=C\begin{array}{c}NH—CH_2OH\\NH—CH_2OH\end{array}$$

$$HO·CH_2·NH·C=O \quad\quad O=C—NH·CH_2OH$$
$$\qquad\qquad|\qquad\qquad + \qquad\qquad|$$
$$NH·CH_2OH \quad HO·CH_2·NH$$

(a) $\longrightarrow$ HO·CH$_2$·NH·C=O $\qquad$ CH$_2$OH
$$\qquad\qquad\qquad\qquad | \qquad\qquad |$$
$$NH·CH_2—N—C—NH·CH_2OH$$
$$\qquad\qquad\qquad\qquad\qquad \| $$
$$\qquad\qquad\qquad\qquad\qquad O$$

(b) $\longrightarrow$ HO·CH$_2$·NH·C=O $\qquad$ O=C—NH·CH$_2$OH
$$\qquad\qquad\qquad\qquad | \qquad\qquad\qquad |$$
$$NH·CH_2·O·CH_2—N$$

(c) $\longrightarrow$ CH$_2$=N—C=O $\;$ O=C—NH$_2$
$$\qquad\qquad\qquad | \qquad\qquad |$$
$$NH·CH_2·N—CH_2OH$$

H$_2$N—C $\overset{N}{\underset{N}{\diagup\diagdown}}$ C—NH$_2$ $\quad$ HOCH$_2$NH—C $\overset{N}{\underset{N·}{\diagup\diagdown}}$ C—NHCH$_2$OH
$$\qquad\qquad\qquad \xrightarrow{+3HCHO}$$
$$NH_2 \qquad\qquad\qquad NHCH_2OH$$

Melamine $\qquad\qquad$ Trimethylol melamine

For adhesive purposes urea-formaldehyde of a sufficiently low degree of polymerization is prepared such that it may be dispersed or dissolved in water to 45–66% concentration. Curiously, they cannot be dissolved completely to a low concentration. The liquid dispersion must be buffered to pH 8 to prevent further polymerization to an insoluble end-product. The buffer should be able to restrain the action of the acid catalyst necessary to effect "curing" of the resin on application to the joint. This is achieved by the incorporation of calcium phosphate and by the use of a salt of a strong acid with a weak base, usually ammonia. The rate of condensation at too acid pH levels is too fast to permit practical utilization of the adhesive.

Melamine resins for adhesives are prepared in solution at pH 10 to prevent condensation during storage. In use, the pH may be lowered to 7, but to effect satisfactory cure, heating is also necessary.

Epoxy Adhesives

The reaction of the epoxy group with reactive hydrogen atoms was described in the section on non-ionics in Chapter 2. Obviously, water itself contains reactive hydrogen atoms, hence a small amount of water, in presence of either hydrogen or hydroxyl ions or "Lewis" acids as catalyst, may initiate the self-polymerization of "epoxy" compounds. Typical Lewis acids are boron trifluoride, aluminium trichloride, ferric chloride and stannic chloride. For example:

$$H_2O + n\, CH_2 \cdot CH_2 \xrightarrow{\text{catalyst}} HO \cdot (CH_2 \cdot CH_2 \cdot O)_n H$$

Condensation of epichlorohydrin with di(p-hydroxyphenyl)-dimethyl methane (bisphenol A) and formation of the epoxy group by elimination of hydrogen chloride in alkali, yields

which, being bi-functional in terms of epoxy groups, is capable of indefinite polymerization. In the above formula, $\bar{n}$ ranges[†] from 0 to

---

[†] $\bar{n}$ is written to indicate an average value.

about 20. If $\bar{n}$ is 2 or more, the product is already solid at room temperature. The liquid products in which $\bar{n} < 2$ are preferred as adhesives although the lower members of the solid products, provided these are soluble in a suitable solvent, may still be used.

Other poly-ols such as glycerin, the novolacs, tetra-p-hydroxy phenyl ethane and so on, may replace bisphenol A yielding adhesives with intermediate molecules of still higher poly-functionality and potential for extensive cross-linking to infusible insoluble end-products. The number of possible epoxy-resins is thus enormous, and no attempt can be made here to cover the entire field.

Either basic or acidic catalysts may be used as "curing" agents to promote completion of cross-linking and formation of a strong joint. Since normally the glue line is thin, heat is also applied to assist the condensation, the heat of reaction passing into the adherends too quickly to be of value in aiding the reaction.

The reactive resins, and many of the catalysts, particularly the amine types, are highly toxic and must not be allowed to come into contact with the skin. The cured resins are without biological effect.

Epoxy resins are particularly versatile adhesives of great strength and toughness. Broken concrete beams may be repaired with "epoxy" with improvement in final strength. They are outstanding in metal–metal bonding and withstand high temperatures. These properties have made possible considerable improvements in aircraft construction. Since they are also excellent insulators, junctions between different metals may be made with "epoxy" without risk of corrosion due to electrochemical couples. They have exceptionally high adhesion to numerous different kinds of surfaces owing to their multiplicity of polar groups, amino, hydroxy, etc., and are therefore extensively used for joining together substances of widely different chemical character such as metals to phenolic plastics. Cohesive strength within the epoxy glue line is generally at least as great as, often much greater than, that of the adherends themselves. Since epoxy glue cures without elimination of water or other small molecules, there are no by-products to get rid of or to cause blistering or other faults. The epoxies are therefore

ideal for bonding impermeable solids. Shrinkage in setting is exceptionally small in the case of the epoxies, hence the final joint is made
with less built-in stresses than with most other adhesives. By incorporation of suitable fillers such as alumina and silica, these residual stresses
may be reduced still further. The cured resin is infusible and insoluble
in all solvents. It maintains shape, without creeping, under prolonged
stressing.

Isocyanate-derived Adhesives

The great reactivity of the isocyanate group with any group carrying
an "active hydrogen atom", as already seen in the case of the epoxy
adhesives, implies that polyisocyanates are of value in the build-up of
polymers of value as adhesives.

$$(a+1)\ HX_1 \cdot R \cdot X_2H + aOCN \cdot Y \cdot NCO$$
$$\longrightarrow HX_1RX_2(OCNH \cdot Y \cdot NH \cdot COX_1RX_2)_aH$$

in which R is any organic radical carrying two reactive groups X
where X may be —OH, —NH$_2$, —NHR, —COOH, —CONH$_2$,
—CONHR, —SH, —SO$_2$OH, —SO$_2$NH$_2$, —SO$_2$NHR, —CSNH$_2$,
—CSNHR, and so on, and Y is any organic radical.

Although solvent swellable cellulose derivatives react readily with
isocyanates, cellulose itself does not in spite of the numerous hydroxyl
groups present. The small amount of water always present in cellulose
reacts preferentially and the disubstituted ureas formed enclose and
entangle the cellulose fibres.

The isocyanates themselves are readily soluble in numerous organic
solvents yielding mobile solutions which readily penetrate porous
adherends, undergoing polymerization within the pores and forming
a structure continuous with that formed in the glue-line exterior
to the pores.

Aromatic polyisocyanates (including di-isocyanates) easily undergo
self-polymerization to three-dimensional structures. The reaction is

illustrated below for a mono-functional compound:

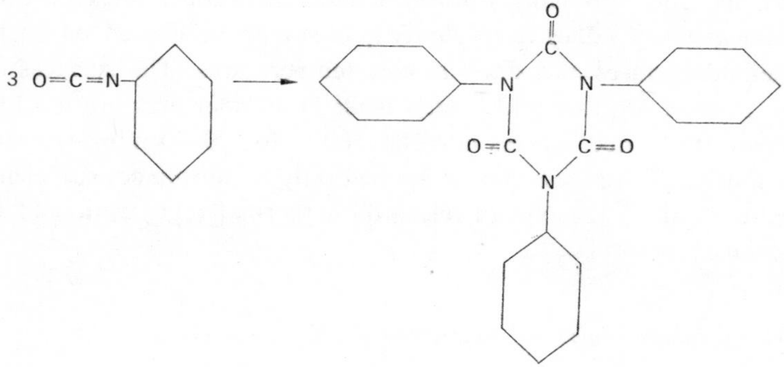

The reaction is promoted by strong heating and by the catalytic action of traces of oxygen, light, tertiary amines and other alkalis, and iron salts.

The capacity of isocyanates to react with so many organic functional groups enables them to form chemical bonds with the surfaces of a wide variety of solid substances in addition to the physical attractions constituting the van der Waals forces. These chemical links greatly strengthen the bonding forces. By reacting with the oxide and/or hydrated oxide layers of many metals, isocyanates may bare the actual metal surface itself. In addition to the van der Waals forces between adhesive and metal, chemical bonds may be introduced by entry of polar groups of the adhesive into coordination linkages with surface metal atoms. Electrical image forces may also be set up.

The great reactivity of the isocyanates also leads to their being powerfully toxic. Stringent precautions are therefore necessary to prevent inhalation of vapour and to provide instant drenching with water of any parts of the skin coming into accidental contact with the liquid.

A particularly convenient method of using isocyanates is to "block" the isocyanate groups by a reagent which is readily removed by heat, regenerating the isocyanate. A suitable "blocking" agent is phenol. The "blocked" product is stable, even in water, at ambient tempera-

tures. The compounded dispersion, which will contain fillers, antioxidants, and optionally rubber or synthetic rubber latex or other
complementary adhesive as desired, is spread or coated on to the
adherend surfaces and dried at low temperatures. The surfaces are
then pressed together and heated while still under pressure until the
adhesive is "cured". Press cures at 180°C for 20–40 min are usual.
Obviously, this process can be applied only to substances sufficiently
porous to allow the phenol released on thermal regeneration of the
isocyanate to escape.

Other Synthetic Polymer Adhesives

It will have become evident to the reader that any organic high
polymer may be adapted to use as an adhesive, provided it can be
reduced to solution form, melted, or formed *in situ*. This is true not
only of the vast variety of vinyl polymers but also of silicones. Some
schools of thought hold that adhesives must be "tailor made" for
particular applications and, arguing from analogy with solutions that
"like attracts like", also hold that the adhesive should be chemically
similar to the substrate. Consideration of the magnitudes of the contribution to the van der Waals forces of the dispersion, orientation,
and induction terms referred to in Chapter 1, and of the principles
of wetting discussed in the same chapter, suggests some modification
of this viewpoint. This point has already been discussed in Chapter 1
at the end of the section on cohesion and adhesion.

In order to get good adhesion between substrate and adhesive, the
first essential is to ensure good *molecular* contact. Solution type adhesives usually do ensure this since the "open time" during which the
solvent is allowed to evaporate is usually ample to allow the dissolved
or dispersed adhesive to adsorb on to the solid surface so that the
individual extended or loosely coiled adhesive molecules may make
many points of attachment to the surface. If the solvent does not, itself,
spontaneously wet the surfaces to be joined, then, provided the surfaces are permeable to the solvent or solvent vapour, forced wetting
must be resorted to, the solution of adhesive being squeezed between

the surfaces in such a manner as to allow the displaced air to escape. The surfaces are then held together until the solvent has evaporated and the adhesive molecules form a continuous structure. The same principle applies to "hot-melt" adhesives. This difficulty arises only when the solvent or hot-melt adhesive is of higher surface energy than the adherend. Once molecular contact has been ensured between the solid adherend–solidified adhesive interfaces, the question of "wetting" or "non-wetting" becomes irrelevant. Much confusion has arisen because, in many practical applications of adhesives, the time necessary to establish the requisite degree of molecular contact across the interface is not allowed.

Reading List

SKEIST, I., *Handbook of Adhesives*, Reinhold, 1962.
HOUWINK, R. and SOLOMON, G., *Adhesion and Adhesives*, Elsevier, 1965.
ALNER, D. J. (ed.), *Aspects of Adhesion*, University of London Press, vols. 1, 2, 3, and 4.
Society of Chemical Industry Symposium, *Adhesion and Adhesives*, 1957.
KNIGHT, J. W., *The Starch Industry*, Pergamon Press, 1969.

CHAPTER 5

Flocculating Agents

Introduction

In the United Kingdom the daily consumption of water per head is some eighty gallons. About half of this is used for domestic purposes and half for industrial purposes. A large part of this nearly 5000 million gallons of water per day is purified before use. A rapidly increasing fraction must be treated after use to yield an effluent of an acceptable standard of clarity and non-toxicity before discharge to rivers. Suspended colloidal matter is rendered capable of settling or filtration by means of flocculating agents, for which purpose many water-soluble high polymers are outstandingly effective.

Some industrial processes in the winning of valuable materials from raw ores start by pulverizing the ore. The valuable mineral is separated from the unwanted rock by flotation methods and in some instances, notably in the recovery of uranium oxide, the required mineral has been so pulverized as to become an almost unfilterable slime. By addition of suitable flocculating agents, to about 0.1% on the mineral content of the slurry, the intractable slime is made easily and rapidly filterable.

Recent developments have led to the production of numerous synthetic polymers for use as flocculating agents, but for many purposes natural or only slightly modified natural products remain the flocculating agents of choice. Reference to the use of a natural polyelectrolyte for the clarification of water is said to occur in Sanskrit literature of about 2000 BC. Work of the Dutch State Mines in the early inter-war years on alkaline and oxidized starches was of the greatest importance. Causticized starch remains a leading industrial flocculant.

The action of decayed organic matter—humus—in improving the crumb structure of, particularly, clay soils, has been recognized in agriculture for some millennia. Investigation of the rational basis of this action by scientific methods at the Rothamstead Experimental Station, Hertfordshire, led to the production of a synthetic resin which exerted the same effect. After some years, this invention was taken up and commercially exploited in the United States. The basic patent of Mowry and Hedrick (1952) discloses improvement in the crumb structure of soils by application of dilute solutions of soluble polyelectrolytes having weight-average molecular weights of at least 10,000. Polymers and copolymers of acrylic acid and maleic acid or various derivatives thereof are the preferred compounds, but the list of monomers which may be used, either alone or in combination, is very large. The preparative details of no less than sixty-three polymers and copolymers are given by way of illustration and numerous examples of the horticulturally favourable effects of the application of the polymers to a soil are shown pictorially. At present, the high cost of these synthetic flocculants is a serious limitation on their exploitation for horticulture and agriculture: a product at a low enough cost would open up a vast potential market.

A symposium on "Improvement of soil structure", published in 1952, stimulated the application of soluble synthetic polymers to a wide range of fields in which flocculation is important.

Kinetics of Coagulation

Dilute Dispersions (i.e. less than 1%)

The particles are assumed to be initially of even particle size randomly distributed and subjected to Brownian motion throughout the period of coagulation. In "rapid" coagulation, particles adhere on colliding as a result of their random movements. In "slow" coagulation, only a fraction of the collisions lead to adhesion of the particles. The assumption that the rate of collision and hence the rate of decrease in the number of particles (doublets, triplets, quadruplets, etc., each counting as "single" particles) with time, is proportional to the square of the instantly present number of particles per unit volume, leads to the equation for rapid coagulation:

$$\frac{1}{N_t} - \frac{1}{N_0} = Kt, \tag{5.1}$$

where N_t is the number of particles after time t, N_0 is the initial number of particles, and K is the coagulation constant.

Calculation of the rate of collision from Fick's diffusion law and the Stokes–Einstein diffusion equation shows that

$$K = \frac{2}{3} \frac{RTa}{\eta N}, \tag{5.2}$$

where R is the gas constant, T is the absolute temperature, η is the viscosity of the medium, and a is a constant of proportionality between the radius of influence of a particle r and its diameter $\mathcal{D}$, i.e.

$$a = \mathcal{D}/r. \tag{5.3}$$

After time t the distribution of the initially monodisperse system into singlets, doublets, triplets, etc., is given by

$$N_i = N_0(t/t_0)^{i-1}/1 + (t/t_0)^{i-1}, \tag{5.4}$$

where N_0 is the initial number of particles (all singlets), N_i is the number of clumps of i particles (i taking the values 1, 2, 3, ..., for singlets, doublets, triplets, ...), and t_0 is a constant, empirically the time for the total number of particles to fall to $0.5 N_0$ and theoretically given by

$$t_0 = 1/4\pi N_0 DR \tag{5.5}$$

in which D is the diffusion coefficient and t is defined by the relation

$$t = 3\eta N/(4RTN_0), \tag{5.6}$$

N being the total number of particles, i.e. $N_1 + N_2 + N_3 + \ldots$ at time t.

In the case of slow coagulation, eqn. (5.4) remains unchanged but eqn. (5.5) becomes

$$t_0 = 1/4\varepsilon\pi N_0 DR, \tag{5.7}$$

where ε is the fraction of collisions which result in adhesion.

It is remarkable that the highly simplified model of Smolukowski, presented above, which ignores such important factors as charges on the particles, ion-atmospheres, van der Waals, and Born forces, etc., which would have rendered the mathematics intractable, yields formulae applicable with a considerable measure of success.

Free Settling

After a time the aggregates will have become large enough to begin to settle under the influence of gravity. If the suspension is sufficiently dilute there will be inappreciable mutual interference of the particles and the conditions of free settling will obtain.

If s is the distance from the base in the vertical axis, the velocity of fall of each particle is given by Stokes's equation

$$v = (\varrho_s - \varrho_0)g\mathcal{D}_e^2/18\eta = z/t, \tag{5.8}$$

where $\mathcal{D}_e$ is the effective diameter of the particle, ϱ_s its density, ϱ_0 the density of the medium, and g is the acceleration due to gravity.

It being assumed that coagulation ceases when settling occurs, the density of the suspension at any point z at time t is given by

$$\Phi(z,\,t) = \varrho_0 + C_0[(\varrho_s - \varrho_0)/\varrho_0 \int_0^{\mathcal{D}_e} F(\mathcal{D}_e)\,d(\mathcal{D}_e)], \tag{5.9}$$

where C_0 is the concentration of the suspension at the moment coagulation ceases and $F(\mathcal{D}_e)$ is its particle size distribution. Differentiation of eqn. (5.8) and application in eqn. (5.9) for a given value of z gives

$$\frac{d\Phi}{dt} = -\mathcal{D}_e F(\mathcal{D}_e)/2t, \tag{5.10}$$

where Φ is the ratio of concentration of particles z at time t to the initial concentration, i.e. $\Phi = C_t/C_0$.

Alternatively, at a given t, the distribution at various levels z is given by

$$dZ/dz = -\mathcal{D}_e F(\mathcal{D}_e)/2z, \qquad (5.11)$$

where z is the ratio of concentration at level z to the initial concentration, i.e. $Z = C_z/C_0$. The complete course of free settling is described by eqns. (5.10) and (5.11).

Concentrated Dispersions (i.e. more than 10%)

Hindered Settling

If the suspension be concentrated, mutual interference between the particles induces them to settle *en masse*, leaving an interface between particles and clear medium. The case is identical with that of liquid flow through a packed bed, the packing density increasing with time. The Kozeny–Carman–Fair–Hatch equation is applicable and the rate of settling is proportional to the Reynolds number of the liquid flowing between the particles. (Reynolds number is a dimensionless number of the form $Lv\varrho/\eta$, where L is a characteristic linear dimension of the system, v the linear velocity of flow, ϱ the density, and η the viscosity of the fluid.)

Steinour gave the equation in the form

$$Q = \{0.123V_s/(1-W_i)^2\}\,\{(t-W_i)^3/(1-t)\}, \qquad (5.12)$$

where Q is the volume of clear liquid appearing in time t, V_s is the volume of solids, and $W_i = \alpha/(1+\alpha)$, where α is a fraction of unit volume of solid to take account of "stagnant" water in the system.

La Mer *et al.* made a very thorough study of the flocculation, settling, and filtration of slurries of finely ground uraniferous phosphate rock in the presence of various flocculating agents. They showed that the three processes are intimately related, and hence only convenience determines which phenomenon is chosen for investigation in any given case. For concentrated dispersions either settling rates or filtration rates may be used. So long as flocculation is incomplete, filtration rates are slow. Although rates of settling may be increasing with increasing degree of flocculation, the remaining suspended fines become increasingly effective in choking the filter, and filtration rate tends to vary inversely with rate of settling. Within the region of complete flocculation, both rate of settling and rate of filtration increase with increasing degree of flocculation, the initial filtration rate becoming proportional to the square of the rate of settling.

Mathematical analysis of the flocculation–settling process led to the equation,

$$t/(h_0-h) = 1/V_0 + t/(h_0-h_f+c) \qquad (5.13)$$

in excellent agreement with experiment. In the equation h_0 is the initial height of the suspended column of solids, h is the height of the settled column after time t, h_f is the final settled height of the suspended matter, c is a factor representing the free water remaining in the network when the solid has fully settled, and V_0 is the initial rate of settling, $-dh/dt$.

The final settled volume is also a measure of the degree of flocculation, but is much less easily determined accurately than the sedimentation rate or filtration rate.

The technique of measuring the "re-filtration" rate, developed by La Mer et al., is simple and reproducible. The whole sample is first filtered to form a filter cake of constant and known thickness. The filtrate is then re-filtered through the filter-bed so formed. The characteristics of the bed are determined by the floc size and the amount of unflocculated material, i.e. they are a function of the degree of flocculation of which the re-filtration rate is therefore an accurate measure. The cake thickness is measured with a travelling microscope. The pressure gradient, ΔP, is monitored by a manometer and controlled to a predetermined value—about 74 cm of mercury is a convenient value.

Combination of the Kozeny–Carman equation with the Langmuir adsorption equation for the adsorption of the flocculating agent on the solid particles, and with the Smolukowski kinetic equations for flocculation, finally produced the relation

$$P_0^{1/2}/(Q_f-Q_0)^{1/8} = A+BP_0 \qquad (5.14)$$

in good agreement with the experimental results. Here P_0 is the initial concentration of flocculating agent, Q_0 is the rate of filtration in the absence of flocculating agent, Q_f is the rate of filtration in the presence of the flocculating agent, and A and B are complex constants for the system under investigation.

$$A = (1+bkW)/bk^{1/8}, \qquad (5.15)$$
$$B = 1/k^{1/8}(1+bkW), \qquad (5.16)$$

where b is the ratio of the adsorption–desorption rate constants for segments of the polymer (flocculating agent) molecule, k is given by sS_0/N, where s is the number of adsorption sites per unit surface area, S_0 is the specific surface area of the solid, N is Avogadro's number, and W is the weight of the solid.

As P_0 is increased, Q_f at first rises, passes through a maximum, and then falls as excess flocculant acts as a dispersant. By equating to zero dQ_f/dP_0 [obtainable from eqn. (5.14)], the optimum initial concentration of flocculant required is given by

$$P_0 \text{ (optimum)} = (1+bkW)^2/b. \qquad (5.17)$$

This is the ratio of the intercept A to the slope B of the linear plot according to eqn. (5.14).

Molecular Mechanism of the Action of Organic Flocculants

Gelatin and other natural high polymers had already long been known as capable, in quite low concentrations, of "protecting" numerous finely divided mineral suspensions against the precipitating action of simple electrolytes, when Schulze, in 1866, pointed out that at still lower concentrations gelatin was as effective as lime or alum (already in common use as flocculants) in causing the rapid sedimentation of clay, and that the addition of minute quantities of gelatin to barium sulphate is able to facilitate its filtration and washing. This latter technique is still good analytical practice. Thus colloidally dissolved high polymers may act as "sensitizers", "protective colloids", or as "flocculants".

The traditional colloid-chemical point of view with respect to sensitization and protective colloid action was that the gelatin, for example, at extremely low concentration, was adsorbed as a monolayer in which the polar groups faced the mineral particle and the non-polar parts of the molecule were turned towards the water. Thus the particle surface was made more hydrophobic and hence tended to be "squeezed out" more intensely by the surrounding water. At rather higher gelatin concentrations a second monolayer was adsorbed having the reverse orientation. The particle surface was now strongly hydrophilic and the tendency to be "squeezed out" strongly diminished or even eliminated.

A different viewpoint was put forward by Overbeek who investigated the sensitizing and protective action of hydrophilic colloids, particularly gum arabic, on a colloidal silver suspension. He thought of the hydrophilic colloid as approximately bolaform and roughly of the same size as the colloidal silver particles (submicroscopic cubic crystals).

If the number of hydrophilic colloid particles was substantially less than the number of colloidal silver particles, the latter, by adsorption, formed an envelope of silver around each hydrophilic particle: the large, hydrophobic, composite particle, because of its greater size, was more readily precipitated than the initial discrete particles. If the number of hydrophilic colloid particles was substantially greater than the number of colloidal silver particles, the former adsorbed on to the silver particles so as to surround each one with an envelope of hydrophilic particles which had no tendency to adhere on collision because of their strong attraction for water molecules. If the numbers of the two kinds of particles were roughly the same, neither could form complete envelopes around all the particles of the other, and the tendency to mutual adsorption could be satisfied most completely

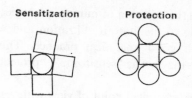

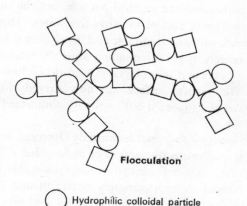

FIG. 5.1. Flocculation.

only by the formation of a network branching in all directions. These concepts are illustrated in Fig. 5.1.

Measurement of adsorption of high polymers on to colloidal dispersions of independently determined specific surface area has shown that much more is adsorbed than corresponds to an extended molecule lying flat on the surface. Increasing chain length of the polymer involved slower adsorption to higher final values. Polymers which are highly effective in concentrated or moderately concentrated suspensions are quite ineffective in highly dilute suspensions of the order of 20–100 ppm by weight, which are often met with in waterworks practice.

These facts strongly suggest that the long-chain polymers are adsorbed in "loops" on the surface of the solid. If the particles are sufficiently close together, a single long-chain molecule may adsorb on to several particles. By increasing the number of "loops" attaching to each particle, the length of polymer molecule between the particles rapidly diminishes, bringing the particles together. If the particles are far apart, a polymer molecule adsorbed on to one may not reach to another. Increasing the number of loops attaching it to the particle reduces still further the chance of attachment to another particle. The Brownian motion of the polymer molecules and particles, given that the attachment of any loop to a particle is strong, will ensure that these sequences will be followed. This concept is illustrated in Fig. 5.2.

Long-chain polymer and concentrated dispersion

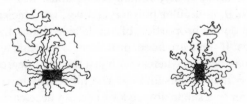

Long-chain polymer and dilute dispersion

FIG. 5.2. Long-chain polymers and particles.

Although most mineral particles are negatively charged as a whole in water (iron oxide is exceptional in being positive), many have heterogeneously charged surfaces—clays, for instance, may have flat, negative faces and thin, positive edges. Anionic polymers are among the most effective flocculants in spite of the overall negative charge of the mineral particles. In some instances, polyacrylamides on quartz particles for example, the flocculating effect, strong on fresh particles, is weak on particles aged in water or on fresh particles in urea solutions instead of in water. Where this is so, adsorption by hydrogen bonding rather than by coulombic attraction between unlike charges is the most probable mode of attachment between particle and polymer. The action of gelatin on clay particles is unaffected by urea; hence hydrogen bonding is less important in this case.

Synthetic Organic Flocculants

The synthetic polymers originally introduced for the improvement of soil crumb structure were of only moderate molecular weights—about 10,000. Development of similar water-soluble polymers as flocculants rapidly led to the introduction of very high molecular weight materials having molecular weights of up to 5,000,000, it having been found that flocculating efficiency tended to increase with increasing molecular weight. Comparison by La Mer *et al.* of a number of flocculants on a silica suspension, at a dosage rate of about 0.4 lb per ton of silica, by the re-filtration method, gave the results shown in Table 5.1.

The effect, already mentioned, of increasing molecular weight, holds only within a series of closely similar compounds. In the case of, for example, very high molecular polyacrylamides, the efficiency is critically dependent on the proportion of amide side-chains hydrolysed to carboxyl groups. If there are none, or if there is too high a proportion, the efficiency may be low, whereas with just the right degree of hydrolysis the efficiency may be outstanding. The nil results in the following table (which refer to a single dosage level) do not necessarily mean that the respective products are ineffective at any dosage level, although this may well be so in some instances. Aged silica suspensions are partic-

TABLE 5.1. FLOCCULATION EFFICIENCIES ON SILICA BY THE RE-FILTRATION
TECHNIQUE

Polymer		Nominal molecular weight	% re-filtration improvement
Trade Name	Type		
Superfloc 16	Polyacrylamide	$3-5 \times 10^6$	325
PAM No. 3	Polyacrylamide	10^5	315
Et-494	Polyethyleneimine	50,000	330
Jaguar	Modified guar gum	?	237
Et-450	Sodium polystyrene sulphonate	?	0
PAM No. 2	Polyacrylamide	$3-5 \times 10^6$	0
Separan NP10	Neutral polyacrylamide	1.5×10^6	0
Separan NP20	Neutral polyacrylamide	3×10^6	0
Separan AP30	Anionic polyacrylamide	$1.5-3 \times 10^6$	0
Lytron X836	Vinyl acetate–maleic anhydride copolymer	?	0
Gantrix AN119	Anionic copolymer of methyl vinyl ether and maleic anhydride	?	0
PAM No. 4	Hydrolysed polyacrylonitrile	0.3×10^6	0
Lytron 810	Styrene–maleic anhydride copolymer	?	0
Flocgel	Causticized starch	?	0

ularly difficult to flocculate, the surfaces becoming almost close-packed with Si–OH groups and hence being strongly water attracting. There is no one "best" flocculant, the most efficient, or the most efficient at a given cost, changing with the nature of the suspended solid, the pH conditions, presence or absence of electrolytes such as sodium chloride or sodium sulphate, and what, if any, inorganic coagulants have been used in conjunction with the organic flocculants, in any particular problem.

In the field of water purification for potable purposes, where the suspended solids content is low, the synthetic polymeric flocculants have, in general, not been found effective. However, it has been claimed that a particular grade of alginic acid is effective at a concentration of

0.2 ppm in substantially reducing the quantity of alum needed as primary coagulant.

The large amounts of clays occurring in coal-washing wash-waters are most cheaply, and very rapidly, settled out by the addition of small amounts of Flocgel—causticized starch. But in the Durham coalfield this is true of one of two neighbouring areas, but not of the other, where satisfactory settling of the suspended solids at an acceptable cost is still a problem. Gelatin and modified glues can give rapid settling with excellent clarity of the supernatant liquor, but not at an acceptable cost.

The use of gelatin (glue and size) for the clarification of colliery wash-waters was proposed by Burrows *et al.* in 1924 and has been repeatedly re-investigated since but has never been applied on the large scale. Difficulties that may be encountered in this kind of investigation are high-lighted by the observation of Needham that, although he could not confirm the efficacy of glue in laboratory studies it was highly effective on the semi-large scale at a dosage rate of about 2 lb per ton of settled solids. A disadvantage of glue is that its properties vary from batch to batch and its efficiency varies markedly with pH, being optimum at acid pH levels. Numerous attempts have been made to improve the efficiency of glue by various chemical treatments, but, as the writer has also found, any improvement obtained is insufficient to make glue products competitive on cost.

Fuoss and Sadek pointed out that even in the most dilute solutions of polyelectrolytes, the charge density of each polyion is considerable. A strong interaction between oppositely charged polyions is therefore to be expected. This is borne out in practice. Turbidimetric titrations of, for example, sodium polyacrylate and polyvinyl butyl pyridinium bromide were possible down to a concentration of 10^{-9} molar in polyion.

The application of this principle to the flocculation of mineral matter in suspension was made by Wadsworth and Cutler who treated kaolinite first with an anionic polymer and then with a cationic polymer. Maximum settling rate was reached at a weight ratio of 3.5 : 1 of cationic to anionic polymer. With this ratio held constant, settling rates continued to increase with increasing total concentration of polymers up to very large concentrations. The order in which the reagents must be added is governed by the response of the solid surface to the pH prevailing.

Long-chain anionic polymers in aqueous solution are highly extended owing to the mutual repulsion of like charges. In the presence of simple electrolytes such as sodium chloride they relax to a less extended, although still open, configuration, since entry of sodium ions into the neighbourhood of the anionic centres diminishes the electrostatic repulsion. However, in the same way the electrostatic barrier to approach of the polymer to negatively charged particle surfaces is also reduced. The overall effect is that the simple electrolyte aids flocculation by the polymer.

In the case of zwitterionic polymers such as gelatin, mutual attraction between the discrete positive and negative charges tends to a coiled configuration of the protein molecule. In the presence of simple electrolytes such as sodium chloride, this mutual attraction is diminished by the approach of sodium ions to the anionic centres; the approach of chloride ions to the cationic centres enables the molecule to take up a more extended configuration, which again aids flocculation.

In general, linear polymers would be expected to be more efficient flocculants than branched polymers of similar molecular weight since they are capable of taking up more extended configurations.

Reading List

DALLA VALLE, J. M., *Proc. Amer. Soc. Civil eng.* (Hydraulic Div.), **82,** 1051–2, 1958.
LA MER, V. K. and HEALY, T. W., *Rev. Pure and Appl. Chem.*, 1963.
MONSANTO CHEMICAL CO., MOWRY, D. T. and HEDRICK, R. M., U.S.P. 2,625,471.
WATER RESEARCH ASSOC., Special Rept. SR4, 1963. Proc. Coagulation Colloquium.
KRAGH, A. M. and LANGSTON, W. B., *J. Colloid Sci.* **17,** 101, 1962.
SYMPOSIUM; Improvement of Soil Stucture, *Soil Science*, **73,** 1952.
BLACK, A. P., BALL, A. I., BLACK, A. L., BOUDET, R. A. and CAMBELL, T. N., *J. Amer. Waterworks Ass.* **51,** 247, 1959.

CHAPTER 6

Dispersing Agents

SOLIDS finely divided into small particles and stirred into a liquid medium tend generally to agglomerate into clumps and settle out. Traditional empirical arts and manufactures, since early antiquity, have used a number of natural organic substances to overcome this tendency in the preparation of inks, paints, cosmetics, ointments, and salves. Gums, resins, glues, and tannins of various kinds were utilized for these purposes. The development of colloid science in the nineteenth century led to the recognition that these substances dissolved to colloidal solutions and hence to their designation as "protective colloids". The ever-growing diversity and sophistication of the products of industry led to demands for "protective colloids" or "dispersants" of standardized composition and performance to a degree unrealizable with natural products.

Softwoods for papermaking contain, in addition to resinous materials, comparable amounts of lignin and cellulose. The lignins are converted to soluble sulphonates in order to recover the cellulose. The by-product lignin sulphonates were found to be excellent dispersing and deflocculating agents well before the turn of the century. The natural tannins have similar properties, and these properties were found in the first decade of this century to be provided also by the synthetic tannins which are condensation products of, for example, sodium naphthalene-2-sulphonate with formaldehyde (to a low order of polymerization). Both lignin sulphonates and synthetic tanning agents are still widely used for making dispersions of insoluble dyestuffs, pigments, pest control products, etc.

In the mid-1920s German chemists discovered that the dispersing or "protective colloid" activity of proteins and gums was enhanced in the

144

presence of organic sulphonic acid salts, especially the lignin sulphonates, in the presence of tri(hydroxyethyl)amine. When the mixture is present in the paste of inorganic or organic water-insoluble pigments undergoing grinding, the improved dispersity increases the colour strength of the pigment.

As early as 1928, German work had shown that not only could exceptional stability be conferred on emulsions and aqueous dispersions by the use of blends of cationic and anionic tensides, but, further, in some instances substances insoluble in water could be brought into clear aqueous solution by such blends. It was specified that the quaternary ammonium cations should contain at least one aromatic substituent, phenyl trimethylammonium and benzyl phenyl dimethylammonium salts being cited as examples. The surface active anions of the blend could be those of Turkey red oils, naphthenic acid salts—aromatic or hydroaromatic sulphonates, alkylaryl or arylalkyl sulphonates, alkylsulphates or sulphonates, and numerous others. It was not essential that the preparations of cationic and anionic tensides should be stoichiometrically equivalent.

Even in the presence of protective colloids, dispersed pigment (vat) dyes had not been successfully applied to the dyeing of heavily woven or twisted linen or mercerized cotton goods. Addition of 0.3 to 10 g per litre of dye bath of polymerized alkylene oxides, not themselves protective colloids, enabled such dyeings to be made with success. These substances strongly retarded the rate of dyeing, ensuring even penetration of the fabric, and greatly enhanced the stability of the dye pastes. Their restraining action on the rate of dyeing renders these polymers valuable for improving the stripping of vat dyes from fabrics by means of strongly reducing hydrosulphite liquor. Polyethylene oxide was the preferred substance. For dye-stripping baths its use at 2–10 g per litre was recommended. This discovery was quickly followed by the further discovery that ethylene oxide condensates of organic hydroxy-compounds other than methanol, ethanol, propanol, butanol, and pentanol were even more effective. The patent applications for these discoveries were filed in September and November respectively of 1930.

It is an interesting commentary on human nature that in spite of the disclosure of the improvement of dyeing by pigments by increasing their stability in dispersion, attempts were being made over the next 20 years to relate the efficiency of detergents to clean cotton goods to their efficiency in "suspending" (i.e. dispersing) finely divided insoluble pigments such as manganese dioxide, carbon, or "vacuum sweeper dust", the latter being assumed more representative of naturally occurring soiling of clothing. The scientist in industry must avoid the twin errors of assuming that his speciality is utterly different from every other speciality and of ploughing so deep a furrow in his own field that he can no longer see over its sides into the other fellows' fields.

"Sulphated oils" are made by treating oils and/or fats, especially those containing a proportion of combined unsaturated and/or hydroxylated fatty acids, with "sulphating agents". The double bond $-CH=CH-$ is converted largely to

$$-CH_2-CH- \atop \qquad\; | \atop \qquad\; O-SO_3H$$

and to a lesser extent to

$$-CH_2-CH- \atop \qquad\; | \atop \qquad\; SO_3H$$

on treatment with sulphuric acid and may form

$$-CH-CH- \atop \; | \qquad | \atop \; Cl \quad SO_3H$$

on treatment with chlorosulphonic acid. When an unsaturated alcohol is sulphated with chlorosulphonic acid, the elements of hydrogen chloride may add across the double bond. Strong sulphating agents such as oleum or S_2O_6 readily sulphonate the α position adjacent to a carboxylic acid or carboxy-ester group. Thus the sulphated oils can be quite complex mixtures, the proportions of the various possible components varying with the molar ratio of reagent to oil, and to the conditions of the reaction, the highly sulphonated oils tending to be more complex than the Turkey red oils. The water-soluble salts of many of the sulphated oils retain the sensitivity of the soaps to hard water.

The Böhme AG of Germany proposed in 1928 to avoid this disability by (a) hydrogenating the glycerides to the fatty alcohols which could then be mono- or di-sulphated, and (b) by conversion of the glycerides to the fatty acid amides, substituted amides, anilides, and the like, followed by sulphating the double bonds and/or hydroxy groups. The modified products are stable to hard water and provide efficient dispersing agents. Similarly, the IG Farbenindustrie (1930) disclosed that α-sulpho fatty acids are readily esterified on the carboxy group (in the absence of mineral acid) and that the water-soluble salts of the α-sulpho esters are stable to hard water.

In 1927 American workers of the du Pont de Nemours reported the preparation of materials which were both less alkaline and more effective dispersants than soaps by heating the higher fatty acids or their glycerides with hydroxylic organic bases in about 20% molar excess. Sulpho-fatty acids could be similarly treated with one or two moles of the bases, of which typical examples are mono-, di-, and tri-ethanolamines, aminopropane diol, and methyl diethanolamine. Later knowledge of the reactions involved shows that with primary and secondary amine alcohols the products must have been mixtures of the organic base salts, ester-amines and amido-alcohols of the fatty acids; with tertiary bases the amido-alcohols of the fatty acids would be absent.

The Society of Chemical Industry in Basle concern disclosed in 1930 that addition of even small quantities of quaternary ammonium compounds derived from unsymmetrically acylated diamines such as oleyl ethylamidodiethyl methyl or hexyl ammonium salts,

$$C_{17}H_{33}CO \cdot N(C_2H_5) \cdot CH_2 \cdot CH_2 \cdot N(alkyl)_3^+ Cl$$

markedly improves the dispersing capacity of such varied materials as soaps, the salts of gall acids (cholic acid; taurocholic acid) and their derivatives, high-molecular sulphonic acids and their salts such as lignin sulphonates, Turkey red oils, albumins, gelatin, glue, saponins, natural and artificial resins, cholesterin derivatives, phosphatides, "gelloses", gums, natural and artificial waxes, wool waxes, solvents, softening agents, and inorganic colloids—a remarkably comprehensive list.

The preparation of aliphatic sulphonates by reacting mineral oils or fatty acids with sulphur trioxide had been described by the IG in 1926. Mineral oils react randomly along the paraffin chain but in the acids the carboxy group activates the α-CH_2 group so that mainly the α-sulpho fatty acids are obtained. Both types of product have excellent dispersing action, especially when used as their salts with amines containing one or more hydroxyl groups such as the ethanolamines. More complex products (1927) are the amine salts of cycloaliphatic or hydroaromatic sulphonic acids, preferably if the molecules also contain, either in the acidic or basic part, alkyl, arylalkyl, cycloalkyl, acyl, or other substituents. It is not essential for optimum dispersing power that the sulphonic acids and the amines should be in stoichiometric proportions. The brevity of the statement tends to hide the extreme breadth of the disclosure.

The du Pont work noted above was followed up by the IG in a broad disclosure of 1929. The reaction was described, of aliphatic or cycloaliphatic carboxylic acids, or their anhydrides, esters or acid halides (in the absence of a metal halide having a condensing action) with secondary nitrogen bases to form amides which could be sulphated or sulphonated or both. The nitrogen bases could be aliphatic, cycloaliphatic, aromatic, aromatic–aliphatic, or heterocyclic. Hydroxyamines were excluded: aromatic groups linked to the nitrogen atom could carry only a sulphonic acid substituent group at most. Suitable amines cited included diethylamine, dibutylamine, ethylhexylamine, methylcetylamine, propylcyclohexylamine, methylbenzylamine, N-ethylaniline, diphenylamine, taurine, morpholine, and piperidine. Carboxy acids derived from animal and vegetable fats or from paraffinic oils and waxes by oxidation, are suitable as the acidic component.

If the acidic component contains a hydroxyl group or a double bond, the elements of sulphuric acid may be added, or if either component contains an aromatic ring not sulphonated before amidification, the ring may be sulphonated after amidification of the carboxyl groups. Amides of lower fatty acids should carry one hydrocarbon group of at least eight carbon atoms attached to the nitrogen atom. Amides of saturated fatty acids may be sulphonated with oleum or sulphur trioxide on the carbon atom α to the amide group.

The soluble salts of the final products may be used alone or in conjunction with other dispersing agents including aromatic, especially polynuclear, sulphonates, hydroxyalkylamines, quaternary ammonium bases, etc. It is claimed that they have a "far-reaching" power for dispersing water-soluble dyestuffs, pigments, fats, waxes, and so on.

A still further disclosure describes the condensation of aliphatic–carboxylic, sulphonic or sulphonated carboxylic acids having a carbon chain of not less than eight carbon atoms, with monohydroxylic monoamines to form the corresponding amino esters or hydroxy-alkylamides or hydroxyalkylsulphonamides. These may be further re-acted with alkylene oxides or with an alkyl halide. Terminal hydroxyl groups may be acetylated or sulphated etc. Terminal amino groups may be alkylated to quaternary ammonium ions. If one of the alkyl groups of the quaternary ion carries a carboxyl group the products are betaine-type taurides and if it carries a sulphonic acid group the pro-ducts are sulpho-betaines—"sultaines". Sultaines are also obtained if the carboxy groups of α-sulpho fatty acids are esterified with quater-nized amino alcohols. All these products are applicable as dispersing agents.

The term "dispersing agent" must seem to embrace a bewilderingly extensive and heterogeneous range of materials. It is therefore useful to look for a unifying concept or concepts.

It is an obvious truism that dispersions may be obtained either by the disintegration of a massive solid material or by the growth of particles by precipitation from molecular solution.

Classic examples of dispersion by disintegration are the ancient development by the Chinese of ink ("Indian" ink) by the patient hand-grinding of graphite in an aqueous solution of gum arabic and the modern manufacture of colloidal graphite by machine-grinding of graphite in an aqueous solution of tannin.

Classic examples of dispersion by precipitation are the preparation of aqueous dispersions of gum mastic in water by rapidly stirring water into which runs a thin stream of mastic in alcohol and the formation of a fine suspension of silver chloride in the titration of a solution of a soluble chloride with a solution of silver nitrate. These examples will serve for a discussion of the stabilization of solid dispersions.

Carbon particles in water exert only weak attractive forces on each other and on water molecules, whereas water molecules exert a powerful attraction for each other. By withdrawing themselves from the small element of volume between two neighbouring carbon particles and thus forcing the particles together, the water molecules reduce the total number of broken water–water bonds.[†] As was emphasized in Chapter 1 in discussion of the forces governing micelle formation in aqueous solutions of tensides, the aggregation of unstabilized particles in water is an expression of the high "cohesive energy density" of water.

Water-soluble gums such as gum arabic are natural polymers of sugar acids (polyuronic acids). Proteins consist of chains of amino acids backed by —CO—NH— groups carrying various proportions of basic, acidic and neutral pendant groups. Natural tannins are mainly chains of highly hydroxylated benzene rings of high molecular weight. As polymers these "protective colloids" are readily adsorbed on the surfaces of solid particles, but are desorbed very slowly, if at all, since, even if the attractive forces per functional group on the polymer molecule are small, the probability that all of the groups attached to the particle will lift off together becomes vanishingly small as the number of attached groups per molecule becomes large. The artificial polymers discussed in Chapter 5 on flocculating agents may also become efficient

[†] Although this was clearly expounded by Edser in 1922 and has been repeatedly emphasized by many careful authors since, the completely erroneous term "hydrophobic bonding" has become fashionable in recent years to describe a precisely similar physical phenomenon.

"protective colloids" if used on dilute dispersions so that the molecules fold up on to individual particles instead of bridging two or more particles. These polymeric materials, whether natural or man-made, will adsorb to carbon and numerous other particles with their less water-attracting groups towards the particle and their more water-attracting groups towards the water. The particles thus become coated with a sheath of water-attracting groups. On close approach of two particles due to random kinetic motion, the energy of translation will be opposed by the energy with which the water-attracting groups tend to draw water into the small element of volume between the particles. The dispersion is said to be stabilized by a "hydration sheath".

Gum mastic, although insoluble in water, carries a number of water-attracting groups which will tend to orientate themselves towards the water as the gum precipitates from its alcoholic solution. In this example the solid is formed in such a manner that it provides its own hydration sheath, and no additional dispersing agent is necessary.

Titration of a dilute solution of a soluble halide with a dilute solution of a soluble silver salt may be carried out either by adding the silver salt to the halide or vice versa. In both processes the insoluble silver halide is first thrown out as a milky dispersion which suddenly aggregates into large flocs in the immediate neighbourhood of the end point. A simple moving boundary electrophoresis experiment shows that in the presence of excess halide ion the suspended particles move towards the positive electrode; in the presence of excess silver ion they move towards the negative electrode. The rate of movement decreases as the stoichiometric end point is approached. In such a case, the particles of silver halide are kept apart by the surface electric charge, negative in the presence of excess halide ion, positive in the presence of excess silver ion. Close to the end point there are too few excess ions of either charge to maintain the surface charge on the particles, and the tendency of the water to form the maximum possible number of water–water bonds asserts itself and forces the particles to aggregate.

In the case of ionic crystal precipitates, among small ions only those of the crystal lattice are "potential determining ions", and it is only in the presence of an excess of these ions that the particles are charged sufficiently for electrostatic repulsion to keep the particles apart. This is no longer true for large ions. For example, colloidal arsenious sulphide may be stabilized in the presence of sufficient of the triphenyl-methane dye Rosaniline (Fuchsin) to reverse the normally negative charge in the colloid. The large dye cation is adsorbed on the surface of the particle sufficiently to provide a stabilizing charge.

The heterogeneous mass of "dispersing agents" are now seen to be those substances which are adsorbable at a solid–solution boundary so as to provide a strong hydration sheath or a surface electric charge sufficient to prevent close approach of particles under random kinetic motion.

Although for the purposes of classification it is customary to regard suspensions as "stabilized by solvation" or "stabilized by electro-static repulsion between like charges", it is well to remember always that nature knows no such arbitrary distinctions: both stabilizing mechanisms are present in greater or lesser degree in every instance. A surface of separation between two distinct phases tends always to produce a separation of electrical charge, whereby the phase of lower dielectric constant tends to be negatively charged. Ionic charges on molecules adsorbed at a surface of separation of two phases not only repel a similarly charged surface, but also attract water molecules.

CHAPTER 7

Commercial Aspects

OFFICIAL statistics of production and sales give some indications for soaps and detergents and for adhesives, either from the Board of Trade publications or, in the case of soaps and detergents, from trade sources. It will be obvious to the reader that the range of possible

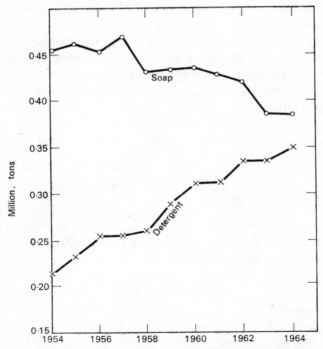

FIG. 7.1. United Kingdom production of soaps and detergents.

adhesives is immense. The entire range of plastics is only part of the field to which mammalian and fish glues and starches are to be added, as is also the large tonnage of inorganic adhesives, particularly silicates, not treated in this book. Statistics of production and usage of adhesives must be particularly difficult to collect and collate.

Board of Trade production data for detergents are on the "as sold" basis, hence the figures for powders include the non-detergent components. Scourers, which may contain a few per cent of tensides, are not included. However, when consumer spending is tabulated, the extraordinary bed-fellow "matches" is included with "soaps and detergents, etc."

The statistics are therefore best regarded as indicators of the order of magnitude of the industry rather than as precision measures. Broad trends, however, will be reasonably well indicated so long as the basis of compilation remains substantially unaltered.

United Kingdom soap production tended to continue to decline slowly and that of detergents to rise after detergent powders had become well established on the market as shown in Fig. 7.1 for the decade 1954–64. The overall totals have increased only slowly in recent years, thus:

	1963	1964	1965	1966	1967
Overall total, soaps and detergents (million tons)	0.747	0.782	0.752	0.761	0.769

Consumer spending is cited by the Board of Trade according to (a) current prices, (b) prices adjusted to a 1958 standard, for matches, soaps, detergents, etc., as follows:

	1960	1961	1962	1963	1964	1965
£M, current	515	543	560	573	609	640
£M, 1958 standard	507	524	523	531	554	560

Note that the 6% rise from 1963 to 1965 compares with a rise of less than 1% in tonnage of soaps and detergents combined over the same period.

The reported values of rubber adhesives sold (both natural and synthetic), including latex adhesives (1964), is about £4M, that of glue and animal size about £4M, and "other adhesives" about £2M.

The number of United Kingdom producers of tensides and detergents is too extensive to be usefully listed here in its entirety, and it would be invidious to list a selected number. Further, mergers, takeovers, and changes in manufacturing policy by individual companies would quickly render an extensive list obsolete. The Society of Chemical Industry publishes an admirable annual *Buyers Guide to the Chemical Industry* which provides ready access to the majority of manufacturers of classified (in many instances individually specified) chemical products. In addition to a general class entry for surface active chemicals, there are entries for wetting, foaming, emulsifying and cleansing chemicals, and for several individual surface active agents, hydrotropes, detergent additives, and organic bases. Similarly, in addition to a general class entry for adhesives there are also special entries for the major different classes of adhesive.

Several trade associations exist whose secretariats will be pleased to help any inquirer in difficulty about sources of supply. These include the following:

Chemical Industries Association (which publishes a directory of the products made by its members).
Adhesive Manufacturers Association.
British National Committee on Surface Active Agents.
The Federation of Gelatin and Glue Manufacturers Ltd.
The Gelatin and Glue Research Association.
The Society of British Soap Makers.

Glossary and Notation

A Area.

A A complex constant in the theory of filtration of settled slurries.

α Coefficient of thermal expansion at constant pressure $(\partial v/\partial T)_P$.

α Polarizability of an atom or molecule.

α, β Angles.

Adduct. Addition complex.

Adsorption. An increase in concentration of a solute at an interface compared with the bulk concentration.

Amphipathic. Term coined by G. S. Hartley from Greek *amphi* = both and *pathein* = be compatible with, to describe molecules in which oil-soluble and water-soluble parts are spatially distinct.

Animal. In casual conversation it is sometimes forgotten that the term "animal" covers many genera of unicellular organisms, soft-bodied invertebrates, insects, crustacea, arachnids, amphibians, fish, reptiles, birds, marsupials, and mammals.

Average. Abbreviations and mathematical terms representing average values are conventionally written with a bar over them, e.g. $\bar{E}$, average potential energy, $\bar{n}$, average number.

b Ratio of adsorption/desorption rate constants for segments of flocculating agent molecules.

B A complex constant in the theory of filtration of settled slurries.

β Coefficient of compressibility at constant temperature $(\partial v/\partial P)_T$.

Biodegradation. Detergents contained in domestic waste finally reach sewage purification plants in which organic matter is degraded by various micro-organisms to innocuous products. Some detergents, especially those with highly branched hydrocarbon chains, resist this degradation process and give rise to serious foam problems in the sewage works, their effluent channels, and rivers. Detergents which are readily degraded are socially desirable and are classed as "biodegradable".

Bolaform. "In the shape of a bolus", i.e. roughly spherical. A large univalent ion in which the charge is central will have an exceptionally small external electrical field.

c A factor representing the free water remaining in the fully-settled solids of a suspension.

c Concentration.

C.m.c. Critical micelle concentration. The lowest concentration at which micelles begin to form.

Cohesive energy density. The function $\Delta H_v/V$, where H_v is the latent heat of evaporation per mole and V is the molar volume.

Collagen. The insoluble protein of the dermis (inner layers of the skin) and tendons.

Country rock. In an ore body, the rock in which the particles of metal, or metallic compound, are embedded.

D Diffusion coefficient.

$\mathcal{D}_e$ Effective particle diameter.

δ The solubility parameter; the square root of the cohesive energy density: liquids with closely similar δ values are likely to be miscible.

Dispersity. Degree of comminution.

$\equiv$ Equivalent to.

E Potential energy.

$\bar{E}$ Average potential energy.

E Gibbs elasticity of a liquid film.

E Young's modulus of elasticity.

ε Fraction of interparticle collisions which result in adhesion of the particles.

ε Intermolecular energy.

η Viscosity.

$(\partial E/\partial V)_T$ The rate of increase of potential energy with increase of volume at constant temperature $= n\Delta H_v/V$.

Enthalpy. Preferred by many writers on thermodynamics to the older term "heat content". The sum of the internal energy of the system and the external work done on the system.

Enzymes. Organic catalysts of biological origin, often specific in their action. Different enzymes are required to break down fats, proteins and carbohydrates to water-soluble fragments.

Epidermis. The outermost layers of the skin. Consists largely of keratin.

f Function of.

f_2 Activity coefficient of the solute.

F Helmholtz free energy.

g The acceleration due to gravity. $= 980.67$ cm per sec per sec.

G Gibbs free energy.

γ Surface tension: interfacial tension.

Γ Adsorption.

Γ_i^j Adsorption of component i on convention j.

h Height of the settled column of solids after time t.

h_f Final settled height of suspended solids.

h_0 Initial height of a suspended column of solids.

H_m Latent heat of melting per mole.

H_{soln} Molar heat of solution.

H_v Latent heat of evaporation per mole.

$°H$ Degrees of hardness of water. English: hardness salts as equivalent of calcium carbonate in grains per gallon, i.e. parts per 70,000. French: calcium carbonate equivalent in parts per 100,000. German: calcium oxide equivalent in parts per 100,000. 56°H German = 100°H French. The English system is mercifully obsolete.

Hydrotrope. A water-soluble substance in the presence of which the aqueous solubility of slightly soluble substances is increased.

I Ionization energy of an atom or molecule.

Isoelectric point (of proteins). The pH at which the numbers of positive and negative charges along the molecule are equal.

ICI. Imperial Chemical Industries, Ltd.

IG. IG Farbenindustrie AG. Powerful German chemical combine broken up after World War II.

k Boltzmann's constant, $= 1.3805 \times 10^{-16}$ erg per molecule per °K.

k A constant.

K An equilibrium constant

°K. Absolute temperature.

Keratin. An insoluble protein. The major component of the outer layers of the skin and of hair, wool, nails, and feathers.

L A characteristic linear dimension.

L Latent heat, e.g., of melting, evaporation or other phase change.

Lewis acid. Any atom, ion or group of atoms which can attach itself to an atom or ion having an unshared pair of electrons.

log. Logarithm to the base 10.

ln. Logarithm to the base e: natural logarithm.

M. Molar.

Micelles. Aggregates of solute molecules.

μ Dipole moment of a molecule.

μ Chemical potential.

$\bar{n}$ Average number of molecules, molecular units, or particles.

N. Mole fraction.

N Avogadro's number $= 6.0226 \times 10^{23}$ per gram mole.

Ossein. A collagen-like protein found in bones.

P Pressure.

P_0 Initial concentration of flocculating agent.

$(\partial P/\partial T)_v$ The rate of increase of pressure with increase in temperature at constant volume.

Pasting property. The property of starch to form an adhesive paste in water instead of a mere gritty suspension.

Phase. Any distinct, homogeneous piece of matter, whether solid, liquid, or gaseous.

Q Volume of clear liquid appearing in time t above a settling slurry.

Q_0 Rate of filtration of slurry in absence of flocculating agent.

Q_f Rate of filtration of slurry in presence of flocculating agent.

r Distance of separation of two molecular centres.

r Radius of a free bubble.

r Radius of influence of a suspended particle.

r_0 Distance of separation of two molecules at equilibrium, i.e. distance at which the sum of the repulsive and attractive potential energies is a minimum.

R The gas constant $= 8.3143$ Joules per gram mole per degree $= 8.3143 \times 10^7$ erg per gram mole per degree $= 1.987$ calories per gram mole per degree.

R Radius of a sessile bubble or drop.

ϱ Density.

s Number of adsorption sites per unit surface area.

S_0 Specific surface area of solid matter of a slurry.

S Entropy: the capacity factor for isothermally unavailable energy.

S_s Excess entropy of the surface.

$\mathcal{S}$ Spreading pressure; spreading coefficient.

SAFE process. Decolorization of an oil using sulphuric acid and fuller's earth.

Sessile. Attached by the base.

Sperm oil. A mixture of esters of higher fatty acids and higher fatty alcohols, mostly unsaturated, obtained from the sperm whale.

Substrate. "The phase lying under".

t Time.

t Width of an annular ridge raised on the substrate by the vertical component of the surface tension of a sessile drop.

T Temperature, °C; absolute temperature, °K.

T Line tension.

T_c Critical consolute temperature.

Tall oil. A mixture of fatty acids and resin-type acids obtained from softwoods used in paper-making.

Tenside. A generic term covering wetting, emulsifying, foaming, dispersing, and cleansing agents. Proposed by an international nomenclature committee to replace "surfactants", "surface active agents", and other terms current in the literature.

°Twaddell. An industrial measure of specific gravity of a liquid. °Tw multiplied by 5 and added to 1000 is the specific gravity relative to water as 1000.

Φ Ratio of concentration of particles at level z at time t_0 to the initial concentration.

v Linear velocity of flow.

V Volume; molar volume.

V_s Volume of solids in a suspension.

V_0 Initial rate of settling of suspended solids, $-dh/dt$.

$(\partial V/\partial T)_P$ Coefficient of thermal expansion at constant pressure.

$(\partial V/\partial P)_T$ Coefficient of compressibility at constant temperature.

W Weight of solid in a slurry.

W_a Work of adhesion between two liquids or between a liquid and a solid.

W_c Work of cohesion of a liquid.

W_i A function of the fraction of unit volume of a solid in a concentrated suspension to take account of stagnant water in the system.

Zwitterion. A doubly-charged molecule carrying both a positive and a negative charge, simultaneously, on different atoms.

The Greek Alphabet

A	α	alpha
B	β	beta
Γ	γ	gamma
Δ	δ	delta
E	ε	epsilon
Z	ζ	zeta
H	η	eta
Θ	θ	theta
I	i	iota
K	$\varkappa$	kappa
Λ	λ	lambda
M	μ	mu
N	ν	nu
Ξ	ξ	xi
O	o	omicron
Π	π	pi
P	ϱ	rho
Σ	σ, ς	sigma
T	τ	tau
Y	υ	upsilon
Φ	ϕ	phi
X	χ	chi
Ψ	ψ	psi
Ω	ω	omega

The Russian Alphabet

А	а	a as in *father*
Б	б	b as in *boy*
В	в	v as in *vice*
Г	г	g as in *good*
Д	д	d as in *did*
Е	е	ye as in *yet*
	(ё)	ya as in *yacht*)
Ж	ж	s = zh as in *pleasure*
З	з	z as in *zero*
И	и	i as in *inn*
Й	й	as e in *pie*
К	к	k as in *kin*
Л	л	l as in *lamp*
М	м	m as in *man*
Н	н	n as in *nine*
О	о	o as in *on*
П	п	p as in *pop*
Р	р	r as in *run*
С	с	s as in *semi*
Т	т	t as in *tea*
У	у	oo as in *boot*
Ф	ф	f as in *fete*
Х	х	ch as in *loch*
Ц	ц	ts as in *its*
Ч	ч	ch as in *churl*
Ш	ш	sh as in *shake*
Щ	щ	shch as in *cash-cheque*
Ъ	ъ	mute (*hard* sign)
Ы	ы	i as in *kiss*, but very short
Ь	ь	mute (*soft* sign)
Э	э	e as in *peg*
Ю	ю	u as in *duke*
Я	я	ya as in *yard*

Size language

tera	10^{12}	T
giga	10^{9}	G
mega	10^{6}	M
kilo	10^{3}	k
hecto	10^{2}	h
deca	10	da
deci	10^{-1}	d
centi	10^{-2}	c
milli	10^{-3}	m
micro	10^{-6}	μ
nano	10^{-9}	n
pico	10^{-12}	p
femto	10^{-15}	f
atto	10^{-18}	a

billion =
10^{12} in English
10^{9} in American

Index